Introduction to Data Science and Machine Learning

Edited by Keshav Sud, Pakize Erdogmus and Seifedine Kadry

Published in London, United Kingdom

IntechOpen

Supporting open minds since 2005

Introduction to Data Science and Machine Learning
http://dx.doi.org/10.5772/intechopen.77469
Edited by Keshav Sud, Pakize Erdogmus and Seifedine Kadry

Contributors
Deanne Larson, Igor Sheremet, Yuriy Zaychenko, Helen Zaychenko, Laura M. Castro, Zlatko Bundalo, Dusanka Bundalo, Robert Nowak, Michał Breiter, Konrad Grochowski, Alicia Martínez-Rebollar, Joaquín Pérez-Ortega, Nelva Nely Almanza-Ortega, Andrea Vega-Villalobos, Rodolfo A. Pazos Rangel, Crispín Zavala-Díaz, Raja Kishor Duggirala, Lucia Duricova Mrazkova, Martin Hromada, Jan Mrazek, Pakize Erdogmus

First published in London, United Kingdom, 2020 by IntechOpen
IntechOpen is the global imprint of INTECHOPEN LIMITED, registered in England and Wales, registration number: 11086078, 7th floor, 10 Lower Thames Street, London, EC3R 6AF, United Kingdom
Printed in Croatia

British Library Cataloguing-in-Publication Data
A catalogue record for this book is available from the British Library

Additional hard and PDF copies can be obtained from orders@intechopen.com

Introduction to Data Science and Machine Learning
Edited by Keshav Sud, Pakize Erdogmus and Seifedine Kadry
p. cm.
Print ISBN 978-1-83880-333-9
Online ISBN 978-1-83880-334-6
eBook (PDF) ISBN 978-1-83880-371-1

For EU product safety concerns:
IN TECH d.o.o., Prolaz Marije Krucifikse Kozulić 3, 51000 Rijeka, Croatia, info@intechopen.com or visit our website at intechopen.com.

Meet the editors

Dr Sud started his research in the field of data science and AI as an undergraduate junior in 2003. He later obtained his MSc and PhD degrees developing a novel mathematical approach for forecasting performance of complex systems from the University of Illinois at Chicago. Dr Sud has worked for fortune 500 companies such as Caterpillar, Volvo, Amazon, and currently holds the role of Senior Manager of Data Science at Keurig Dr Pepper Inc., where he is leading the development of recommendation engines. Dr Sud is a well-recognized expert in the field of data science and AI with first authorship of numerous scholarly articles in internationally recognized professional publications. Dr Sud serves on the editorial board for international journals such as JMES (Sage Publishing, UK) and STM Journal India, and as guest editor for many peer-reviewed journals such as SAE and Hindawi.

Pakize Erdogmus was born in Erzurum/Turkey in 1972. She received her BSc degree in electronic and communication engineering from Yildiz Technical University, Kocaeli Engineering Faculty in 1993. She also received her MSc degree in computer sciences and her PhD degree in numerical methods. She has studied continuous optimization problems, linear programming and revised simplex method, course scheduling, and discrete optimization problem. From 2003 to 2010, she worked as an Assistant Professor in the Duzce University, Technical Education Faculty. From 2010 to 2013 she worked as an Assistant Professor in Duzce University, Engineering Faculty, Computer Engineering Department. From 2013 to 2019, she worked as an Associate Professor in the same department. And from 2019 October, she is working as Professor in the same department. Her research interests are nature-inspired optimization algorithms, machine learning, and signal and image processing.

Seifedine Kadry, PhD, is an Associate Professor of data science at Beirut Arab University, Lebanon. He serves as Editor-in-Chief of the International Journal on Computer and Communications Networks, Computational Intelligence and Data Analytics. He has worked as Head of Software Support and Analysis Unit of First National Bank where he designed and implemented the data warehouse and business intelligence. In addition, he has published several books and is the author of more than 100 papers on applied math, computer science, and stochastic systems in peer-reviewed journals. At present his research focuses on smart learning in smart cities, social network analysis, and E-systems. He received a PhD in computing 2007 from the Blaise Pascal University (Clermont-II) -Clermont-Ferrand in France. He is a Distinguish Speaker of the IEEE computer society.

Contents

Preface

The field of data science has seen remarkable growth in the last decade. It is also the foundation of machine learning, another field experiencing rapid growth and a class of AI upon which a multitude of products and applications are being developed. Using open-source programming and code sharing, developers are able to build completely new applications starting with small pieces of freely available code and building on this using free development tools, which has significantly reduced application development times and costs. As a result, data science applications exist in almost every industry from technology to healthcare, and even government and defense. Inspired by the growth of data science applications built using open-source programming, this book provides a collection of work from international experts and researchers, discusses data science architecture and development best practices, popular algorithms, and applications of data science.

As the chief editor of this book together with my co-editors, I am proud to share "Introduction to Data Science and Machine Learning", a book with the goal to provide its readers with a deep understanding of data science application development and the benefits of using open-source programming. This book is divided into four sections: the first section contains an introduction to the book, the second covers the field of data science, software development, and open-source based embedded hardware; the third section covers algorithms that are the decision engines for data science applications; and the final section brings together the concepts shared in the first two sections and provides several examples of data science applications.

This book was only possible with the hard work of many individuals. I would like to thank all the contributors for sharing their research with our readers, I would like to thank my esteemed co-editors Professor Seifedine Kadry (Associate Professor at Beirut Arab University, Lebanon) and Professor Pakize Erdogmus (Full Professor at Duzce University, Turkey) for contributing, editing, and preparing the content, and finally I would like to thank the IntechOpen team for managing the entire project.

Dr. Keshav Sud
The University of Illinois at Chicago,
Senior Manager Data Science at Keurig Dr Pepper,
Ex-Scientist at Amazon Robotics

Pakize Erdogmus
Professor,
Duzce University,
Turkey

Seifedine Kadry
Professor,
Beirut Arab University,
Lebanon

Section 1

Introduction

Chapter 1

Introductory Chapter: Clustering with Nature-Inspired Optimization Algorithms

Pakize Erdogmus and Fatih Kayaalp

1. Introduction

Humanity has been inspired from nature along its evolution, since ancient times. Each lively being has its own rules and magnificent knowledge. This capability of all lively beings gives inspiration to the human being in order to find solutions to the problems that he/she faces.

Most of the engineering designs have been inspired by nature. With the design of high-speed trains, the problem was "boom effect," created by the trains, when entering the tunnel. This noise was because of the air pressure created on the front side of the train. This problem was solved with an excellent nature design, with kingfisher beak [1].

For more than half a century, algorithms have also been using inspirations from nature for computing and solving the problems related to computer science. The first optimization algorithm mimicking nature was genetic algorithm (GA). Genetic algorithm used the selection, mutation and crossover, finding the diverse solutions to complex problems [2]. Today, we have very powerful algorithms inspired by nature to optimize the problems.

Particle swarm optimization (PSO) is another population-based algorithm inspired by nature. Improved by James Kennedy and Russell C. Eberhart, PSO simulates the bird flocking and fish schooling foraging behaviors for the solution of a continuous optimization problem [3].

Biogeography-based optimization (BBO) algorithm is an evolutionary algorithm that simulates the formation of the biogeographies. BBO, improved by Simon [4], simulates the habitants' immigration or emigration behaviors according to the suitability of the habitat for them.

Gray wolf optimizer (GWO) is also a nature-inspired population-based optimization algorithm originally proposed for the solution of the continuous optimization problems. GWO simulates the hunting behaviors of the gray wolves [5].

Optimization is a kind of programming, solving several problems including function minimization, clustering and feature selection. Clustering is an unsupervised machine learning method that groups the entities with a given number of categories according to their similarities. It is certain that clustering must maximize the similarities of the objects inside the same groups and also maximize the dissimilarity among the other groups' objects. Clustering can be defined as an optimization problem with this perspective. A classical example of clustering is given in **Figure 1**.

Clustering is a very common technique used for data analysis especially for the applications of summarization, abstracting the data and segmentation [6].

IntechOpen

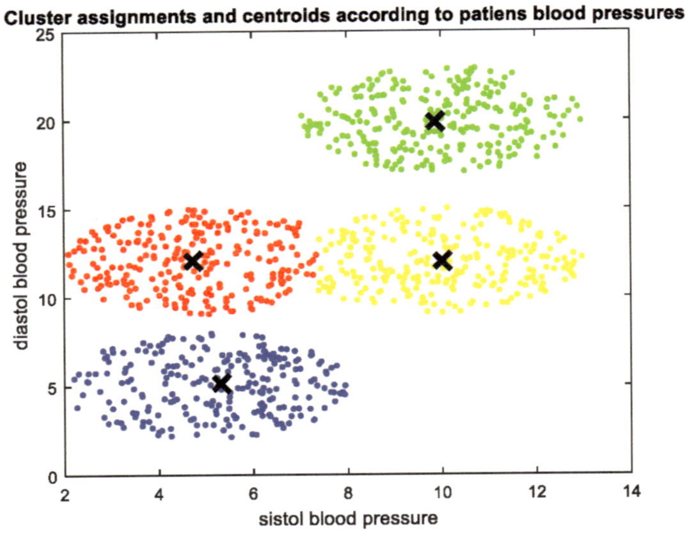

Figure 1.
Patient clusters according to their systole and diastole blood pressures.

A very common application of clustering is data analysis [7]. Cluster centroids give brief information for the attributes of each cluster. This knowledge is used for information discovery and general classification. Another application of clustering is collaborative filtering [8]. The users, grouped in the same cluster, are accepted similar likes and dislikes. Data and image segmentation are another application of clustering [9].

Today, clustering is commonly used for biological data [10], medical data [11], social network [12, 13] and wireless sensor network data [14] and big data [15] for different kinds of applications stated above.

In this chapter, the reader will learn how he/she can apply optimization algorithms for clustering problems. In the next section, the clustering is defined as an optimization problem. Nature-inspired optimization algorithms, genetic algorithm, particle swarm optimization algorithm, biogeography-based optimization algorithm and gray wolf optimization algorithm have been explained in Section 3. Clustering with nature-inspired algorithms has been studied for a very basic and popular dataset given in Section 4. And the results have been submitted in Section 5.

2. Clustering as an optimization problem

Clustering is grouping the data into the clusters according to their similarities. Similarity is defined mathematically with a measure. The more the attributes of two data are near to each other, the less distance is between data. Namely, distance is inversely proportional to similarity. Different distance measures have been defined for clustering. Euclidean, Manhattan, Mahalanobis and Minkowski are the most [16, 17] used distance metrics. The most popular metric for continuous features is the Euclidean distance [18]. Euclidean distance is used while clusters are compact and the dimension of the data is low. For large dimension, Minkowski distance is preferred.

In this chapter, the objective function is defined based on Euclidean distance metric for comparison. Let us assume the two data points in D dimension space, X and Y. The distance between X and Y is calculated with Euclid and pth order of Minkowski distance, as given in Eqs. (1) and (2), respectively.

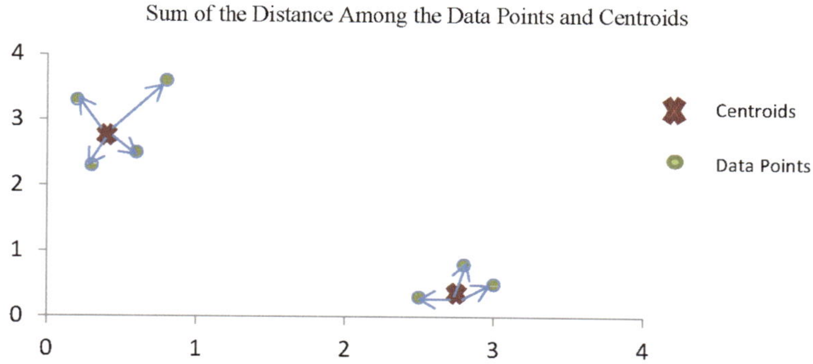

Figure 2.
Minimum distance for optimum centroid positions.

$$De = \sqrt{\sum_{i=1}^{D} (X_i - Y_i)^2} \quad (1)$$

$$Dm = \sqrt[p]{\sum_{i=1}^{D} |X_i - Y_i|^p} \quad (2)$$

The object of clustering is to assign data to the clusters that minimize the sum of the distances from the data to centroids of the clusters. So the objective function with Euclid distance is defined as given in Eq. (3).

$$F_{obj} = \sqrt{\sum_{j=1}^{K} \sum_{k=1}^{D} (X_{ik} - C_{jk})^2} \text{ for } \forall X_i \in C_j \quad (3)$$

where N presents the number of data; K presents the number of fully separated clusters; D presents the number of dimension of data; X_{ik} presents the ith data kth feature; C_{jk} 1, 2, … , K presents the center of the cluster j of kth feature.

The positions of centroids are independent variables. So if applied data dimension is D and the number of cluster is K, the number of independent variables for objective function is KXD. Namely, K centroid positions with D dimension are the independent variables of objective function. The objective is to find centroid positions that minimize the distance. The calculation of the objective function is shown schematically in **Figure 2**.

3. Nature-inspired optimization algorithms

In this chapter, some of the most cited and successful algorithms have been selected for comparing the clustering performances. Genetic algorithm, particle swarm optimization algorithm, biogeography-based optimization algorithm and gray wolf optimization algorithm have been selected. These algorithms have common features. All of them run a group of solutions.

Each solution is called as individual, particle, island and wolf, respectively. In this chapter, the number of solutions is given as S. Each solution has independent variables. So, the number of independent variables is called V, which is equal to KXD for clustering problem. The clustering problem is handled as an unconstrained optimization problem in this chapter as given in Eq. (4).

$$F_{obj} = f(C_1, C_2, \ldots. C_V) \quad (4)$$

Algorithm	Solution	Group of solutions	Parameters
Genetic algorithm	Chromosome	Population	Crossover rate
	Individual		Mutation rate
Particle swarm optimization	Particle	Swarm	Inertia weight/constriction factor
			Social and cognitive parameters
Gray wolf optimization	Wolf	Group	A: linearly decreased from 2 to 0
			C: random number between 0 and 2
Biogeography-based optimization	Habitat	Ecosystem	Migration parameters
			Mutation rate

Table 1.
Population-based optimization algorithms and their naming for common terms of optimization.

In the problem, the initial cluster centroids as independent variables are assigned randomly between the lower and upper values of data. After the independent variables are created randomly, the objective function value is calculated as given in Eqs. (3) and (4). Attaining for S initial solution, V random cluster centroids for each S solution are assigned. Solutions are called F_{obj1}, F_{obj2}, … , F_{objS}. The optimization algorithm starts with these initial solutions and evaluates and improves the solutions, until the stopping condition is true. From one iteration to the other, the algorithm converges to the best solution.

Before the algorithms are explained in detail, the general properties of population (swarm)-based optimization algorithms and specific namings are listed in **Table 1**.

3.1 Genetic algorithm

Genetic algorithm is one of the most studied and powerful optimization algorithms, used for the solution of both combinatorial and continuous optimization problems. The main idea behind GA is "survival of the fittest." So, the algorithm is based on the evolution of the individuals from one generation to the next.

After the optimization problem is modeled and its independent variables, constraints and objective function are specified, genetic algorithm parameters are adjusted for the problem. After the algorithm starts with an initial population, fitness value of each individual in the population is calculated. The selection process for the next generation is realized with some selection methods in such a way that best individuals have more chance than the worse ones. Tournament selection, roulette wheel selection and rank selection are some of the selection methods [2].

After selection of the parents, crossover is applied for the parents. In GA, in reverse to the real evolution, the number of population is constant, the number of child is selected as two, and the best individuals are copied like genetic cloning. Crossover operation is applied with a crossover rate. Zero crossover rate means the children will be the copy of their parents, one crossover rate means the children will be completely different from their parents. After crossover, mutation is applied with a very low mutation rate. Mutation is the permanent changes in genes, in order not to get trapped in local minimum. After the new generation is attained, the fittest ones are selected among the latest population. And algorithm stops after a number of generations. Stopping condition is generally selected as maximum number of generation.

The pseudocode of GA is given in **Table 2**.

```
Generation = 1
Specify max_generation value
Generate S initial solution
While Generation < max_generation
        Evaluate Fitness function values
        Select best solutions for the next generation
        Apply crossover for selected individuals
        Apply mutation for selected individuals
        New Population = selected individuals
        Generation = Generation + 1;
end
```

Table 2.
The pseudocode of GA.

3.2 Particle swarm optimization

PSO is another most studied optimization algorithm, used for the solution of continuous optimization problems introduced by Eberhart and Kennedy [19]. The bird flocking or fish schooling moves in a multidimensional space in such a way that they find the food in a shortest path. The main idea behind the PSO is the behavior of the particles in a swarm. Each particle has a position in a multidimensional space, and they exchange information among them. The particles move in a space using social and cognitive information. When the algorithm stops, the best position has been found.

The algorithm starts after initial positions and initial velocities of particles have been assigned. The dimension size of the particle position in PSO is the number of independent variables. Fitness value of each particle in the swarm has been calculated. The particles update their velocities according to velocity formula. Although two different velocity formulas have been defined, there are two main parameters in both formulas, representing the social and cognitive behaviors of the particles. In swarm, particles update their velocities according to both the best position in the swarm and to their best. In this way, from one iteration to the other, PSO converges the optimum solution of the problems. PSO is the fast convergent optimization algorithm and requires less memory and there are a few parameters to adapt. In the first velocity formula, there was no inertia weight [19]. Inertia weight is introduced by Shi and Eberhart [20]. Inertia weight balances the algorithm's local and global search ability. Inertia weight specifies the percentage of contribution of previous velocity to its current velocity. The velocity and position formulas for PSO are given in Eqs. (5) and (6), respectively.

$$v_i^{k+1} = wv_i^k + c_1 rand\left(p_{besti} - x_i^k\right) + c_2 rand\left(g_{best} - x_i^k\right) \tag{5}$$

$$x_i^{k+1} = x_i^k + v_i^{k+1} \tag{6}$$

where w presents the inertia weight, v_i^k presents the velocity of ith particle for kth iteration, x_i^k presents the position of ith particle for kth iteration, p_{besti} presents the local best solution of ith particle, g_{best} presents the global best solution, rand() presents uniform random number, and c_1 and c_2 present the cognitive and social parameters.

Constriction factor (K) is used by Clerc [21]. Constriction factor assures the convergence of the PSO. The velocity and position formulas with constriction factor for PSO are given in Eqs. (7)–(9), respectively.

$$v_i^{k+1} = K\left(v_i^k + \varphi_1 rand\left(p_{besti}^{\ k} - x_i^k\right) + \varphi_2 rand\left(g_{best} - x_i^k\right)\right) \tag{7}$$

$$K = \frac{2}{\left|2-\varphi-\sqrt{\varphi^2-4\varphi}\right|}\varphi = \varphi_1 + \varphi_2\varphi > 4 \quad (8)$$

$$x_i^{k+1} = x_i^k + v_i^{k+1} \quad (9)$$

where φ_1 and φ_2 are individual and social parameters. The pseudocode of PSO is given in **Table 3**.

3.3 Biogeography-based optimization

BBO applies biogeography mathematical foundations to solve the optimization problems. Biogeography observes the distribution of species in geographic space and tries to find the reason of the biodiversities in geography. Species migrates from one habitat to the other, trying to find the most suitable habitat. So if this biogeographic movement is simulated well, it can be applicable to solve an optimization problem. Geographical areas that are suitable for biological species are said to have a high habitat suitability index (HSI) [4]. The features, such as land area, temperature and rainfall show the suitability of the biogeography and called as suitability index variables (SIVs) independent variables of the optimization problem and HSI represents the fitness function. Species living in a geography that has high HSI emigrates to nearby habitats, which has low species, since this biogeography is already nearly saturated. BBO has been used for clustering in some studies [22, 23]. As seen in **Table 3**, since BBO algorithm uses three loops, BBO runs slower than the other algorithms like PSO and GWO. So some strategies have been used in the studies that make BBO run faster. The pseudocode of BBO is given in **Table 4**.

3.4 Gray wolf optimizer

Gray wolf optimizer, a population-based, nature-inspired algorithm, simulates the hunting behaviors of gray wolves [5]. Gray wolves live in groups, and there is a hierarchy among them. Their hunting strategy has three steps: encircling the prey, circling the prey and hunting the prey. This process is adapted for the optimization problem solution. The wolves move in d-dimensional space in order to search their prey. The position of the wolves presents d the independent variables. After they find the prey, they encircle their preys and lastly they hunt. Encircling behavior presents the converging of the solution and hunting presents the optimum point. The algorithms start with the creation of the initial positions of the wolves. The positions are evaluated with the fitness function. Since there is no knowledge about the position of the prey in problem, the best three solutions are selected, in order to

```
Create P initial particle position
Do
    For I = 1:P
        Evaluate Fitness function values
        If fitness(P_i) < Pbest(I)
Pbest(I) = P_i
        end
        If fitness(P_i) < Gbest
Gbest = P_i
    end
  end
Until stopping condition is true
```

Table 3.
The pseudocode of PSO.

```
Initialize the SIVs of N habitat
Calculate HSI values of each habitat
Sort them and find best HSI
for i = 1 to maximum_iteration
        for i = 1:N
            for k = 1:dim
CandidateNewHabitat = Habitat
                Select Source Habitat
                Apply migration with a probability
                Apply mutation with a probability
        end
        end
        Calculate HSI values for new habitat
        Sort CandidateNew Habitat
        Create NewHabitat from Habitat bests and CandidateNewHabitat
        Update Best Solution Ever Found
End
```

Table 4.
The pseudocode of BBO.

```
Initialize the Gray Wolf Population
Initialize parameter A,a,C
Calculate each wolf fitness value
Specify first,second and third best solutions
while (t < max_iteration)
     for each wolf
         Update the position
end
    Update a,A,C
    Update fitness of each wolf
    Update first, second and third best solutions
    t = t + 1;
end
```

Table 5.
The pseudocode of GWO.

update the next positions of the wolves. Instead of saving only one global best solution in memory, GWO saves three best solutions. This property makes the algorithm powerful for global best finding. GWO is applied successfully in feature selection [24], training multilayer perceptrons [25] and clustering [26–28].

The pseudocode of GWO is given in **Table 5**.

4. Clustering performances of the algorithms

As stated in another chapter, the object of clustering is to assign data to the clusters that minimize the sum of the distances from the data to centroids of the clusters. So the **objective function** value is accepted evolution metric for clustering. In **Tables 8** and **9,** F_{obj} column is given for the other algorithms' clustering performance comparison. In this section, the clustering performances of the algorithms have been compared for IRIS dataset. The parameters, used in the simulation, have been given in **Table 6**.

The benchmark dataset is quite well-known as IRIS dataset [29]. The dataset has four attributes and three class as given in **Table 7**.

All simulations have been implemented on a personal computer with Intel Core Duo 3.0 GHz and 8 GB RAM. Each algorithm has been simulated 30 times and

Parameters	PSO	GA	GWO	BBO
Population size	5–30	5–30	5–30	5–30
Maximum iteration	100–500	100–500	100–500	100–500
Crossover rate	—	0.8	—	—
Mutation rate	—	0.01	—	—
Self-adjustment rate	1.49	—	—	—
Social adjustment rate	1.49	—	—	—
Inertia weight	1.1	—	—	—
a	—	—	$2 \rightarrow 0$	
Habitat modification probability	—	—	—	1
Mutation probability	—	—	—	0.005
Elitism rate	—	—	—	0.05
Immigration probability	—	—	—	[0–1]

Table 6.
Parameters of the optimization algorithms.

Attributes	Classes
Sepal length in cm	*Iris setosa*
Sepal width in cm	*Iris versicolour*
Petal length in cm	
Petal width in cm	*Iris virginica*

Table 7.
The attributes and classes of the IRIS dataset.

results have been saved. The average, best and worst clustering performances have been calculated from 30 runs. As it has been seen, population size and the maximum iteration number are two important parameters, in order to get best solutions in the nature-inspired optimization algorithms. So in order to get optimum values, two parameters must be selected in such a way that both solution time and optimum value must be optimized. With this aim, firstly population size is selected as constant and the number of maximum iteration is changed as 100, 200, 300, 400 and 500. But only the results for iteration number 100, 200 and 300 have been shown in **Table 8**, so that the rows of the table aren't too many.

Secondly, population size is changed as 10, 20 and 30, while maximum iteration number is constant and equal to 200. The results have been shown in **Table 9**.

As it has been seen in **Table 8**, the minimum objective value for iteration number = 100 and the population size = 5 belongs to GWO. PSO is the second best algorithm for clustering. These minimum values found with GWO and PSO are not far from the minimum distance found with k-means. But the average values are quite far from the minimum objective value. So it can be said that both the population size and iteration number are not enough for finding near optimum values for clustering problems [30]. So the algorithms are not stable for these parameters. Average objective values for iteration numbers have been shown in **Figure 3**.

As it has been seen in **Figure 3**, PSO and GWO are fast convergent algorithms. But GA and BBO are also showing similar characteristics, since they have a lot of parameters like mutation rate.

Algorithm	Iteration number		Time (s)	F_{obj}
K-means	100	Average	0,014958	85,24,339
		Min	0,002778	78,85,144
		Max	0,195,176	142,7541
GA		Average	12,675	890,8601
		Min	1,112,884	182,4538
		Max	1,139,856	2071,767
PSO		Average	1,373,846	268,877
		Min	1,338,447	97,33,318
		Max	1,335,599	681,3707
BBO		Average	2,637,282	749,6682
		Min	1,331,544	184,2395
		Max	2,717,847	2044,928
GWO		Average	1,365,047	243,7228
		Min	1,317,441	83,71,005
		Max	1,440,723	692,7803
GA	200	Average	2,693,518	771,0638
		Min	3,119,822	204,5146
		Max	2,432,806	2390,792
PSO		Average	343,235	190,5667
		Min	3,455,119	80,04687
		Max	1,558,096	681,3707
BBO		Average	6,591,394	769,89
		Min	4,468,819	125,669
		Max	5,675,291	1717,338
GWO		Average	3,352,038	204,6926
		Min	3,352,038	81,44,311
		Max	6,069,114	774,6732
GA	300	Average	3,190,949	876,8533
		Min	2,670,085	290,7814
		Max	0,584,012	2148,225
PSO		Average	3,582,182	301,8806
		Min	4,470,383	80,00451
		Max	111,881	862,6507
BBO		Average	7,955,762	674,7325
		Min	8,167,651	198,4093
		Max	8,294,291	1412,369
GWO		Average	3,901,675	169,6811
		Min	384,877	79,77,414
		Max	3,964,712	681,3854

Table 8.
IRIS clustering results of the algorithms for population size = 5.

Algorithm	Pop size		Time (s)	F_{obj}
K-means	10	Average	0,014958	85,24,339
		Min	0,002778	78,85,144
		Max	0,195,176	142,7541
GA		Average	5,298,525	316,0649
		Min	5,094,453	99,18,655
		Max	5,589,196	931,0285
PSO		Average	552,843	126,6415
		Min	5,541,587	78,86,165
		Max	5,511,388	176,8169
BBO		Average	11,19,563	295,1141
		Min	10,73,245	99,54,019
		Max	13,60,106	837,8205
GWO		Average	6,707,655	141,2761
		Min	5,483,527	79,59,426
		Max	908,555	226,1199
GA	20	Average	11,11,173	124,9446
		Min	10,03661	78,8596
		Max	13,36,941	164,6155
PSO		Average	8,826,531	123,2611
		Min	5,475,998	78,85,145
		Max	13,67,359	152,348
BBO		Average	21,39,521	162,8367
		Min	21,10,746	79,57,156
		Max	22,08803	227,7017
GWO		Average	10,61,837	136,0126
		Min	10,43,832	78,90,892
		Max	11,68,964	237,9805
GA	30	Average	17,91,835	114,2972
		Min	14,41,476	78,85,246
		Max	23,24,536	152,3933
PSO		Average	13,68,435	105,8372
		Min	7,912,835	78,85,144
		Max	18,75,625	152,348
BBO		Average	37,57,738	106,0112
		Min	32,82,001	78,85,754
		Max	50,14,788	152,46
GWO		Average	17,19,025	124,1055
		Min	16,05447	78,988
		Max	19,14,221	153,6034

Table 9.
IRIS clustering results of the algorithms for iteration number = 200.

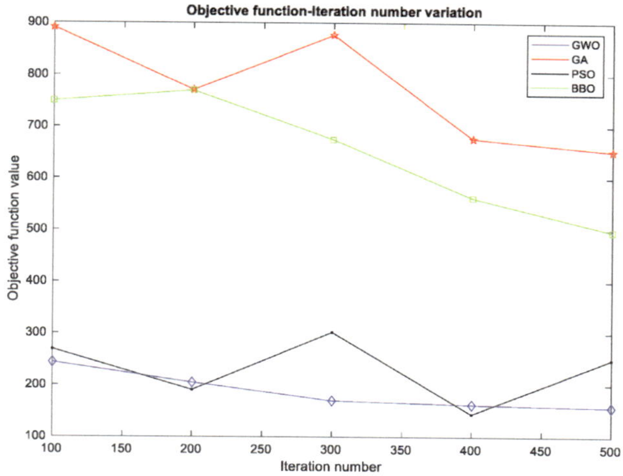

Figure 3.
Average objective values for iteration numbers = 100, 200, 300, 400 and 500.

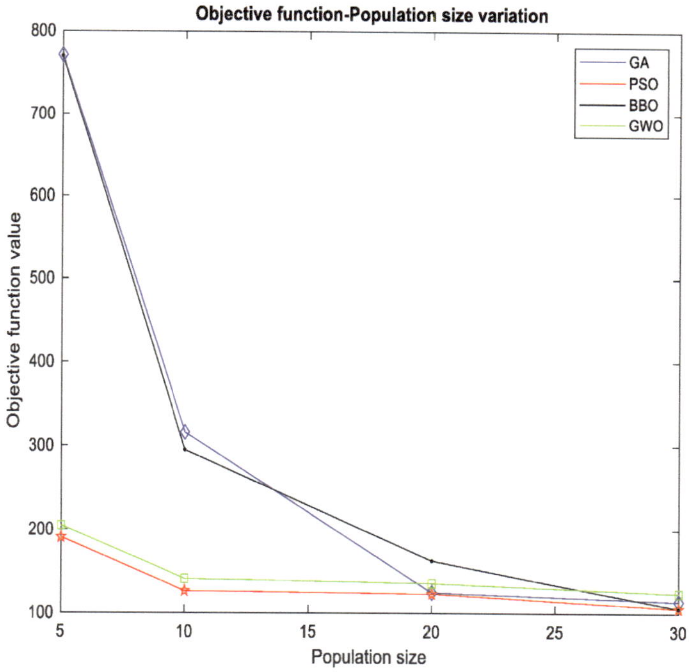

Figure 4.
Average objective function value for population size = 5, 10, 20 and 30.

PSO is the best algorithm for clustering the data with minimum distance from centroid to each data for iteration number = 200. GWO is the second best algorithm for data clustering.

GWO is the best algorithm for clustering the data with minimum distance from centroid to each data for iteration number = 300 and PSO is the second. GWO and PSO are more stable than BBO and GA.

But it has been seen that this population size (population size = 5) is not enough for the algorithms' convergence to the minimum distance for clustering.

As it has been seen in **Table 9**, the best stable values belong to PSO and GWO. But four of the algorithms are working well under the conditions population

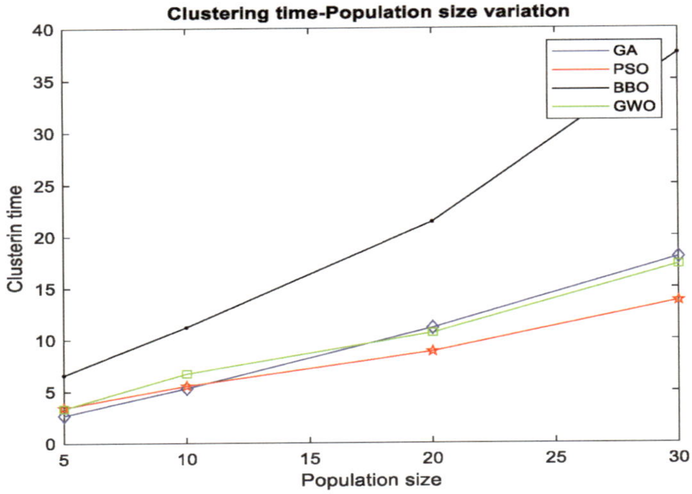

Figure 5.
Average clustering time values for population size = 5, 10, 20 and 30.

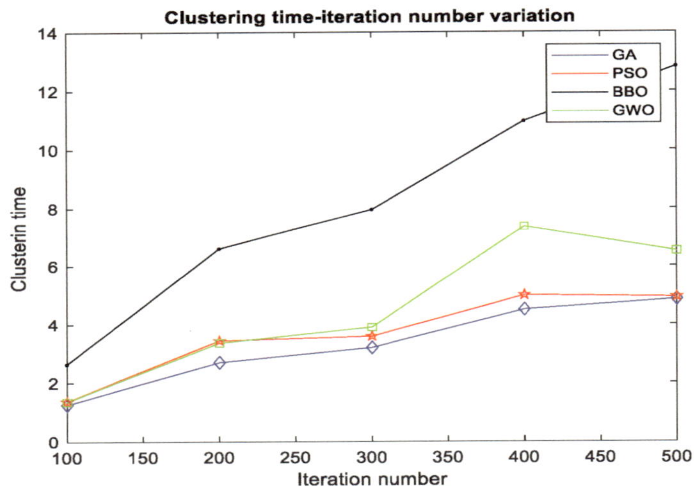

Figure 6.
Average clustering time values for iteration number = 100–500.

size = 30 and iteration number = 200. But clustering time is increasing with the number of population size. Clustering time and objective function value for population size = 5, 10, 20 and 30 have been shown in **Figures 4** and **5**, respectively.

As it has been seen in **Figure 4**, PSO and GWO can produce near optimal solutions for small population size.

However, BBO and GA require many people to effectively operate their mechanisms, such as crossing and mutation. GA and BBO catch the performances of the PSO and BBO after the population size is more than 20.

As it has been seen in **Figure 5**, BBO clustering time is highly increasing with the population size. Solution time for PSO, GWO and GA is changing less, while the population size increase.

Lastly, clustering time variation with iteration number has been shown in **Figure 6**. As it has been seen, GA and PSO clustering time are robust than BBO and GWO, depending on the number of iterations.

As an example, GWO convergence curves for 30 runs have been shown in **Figure 7**.

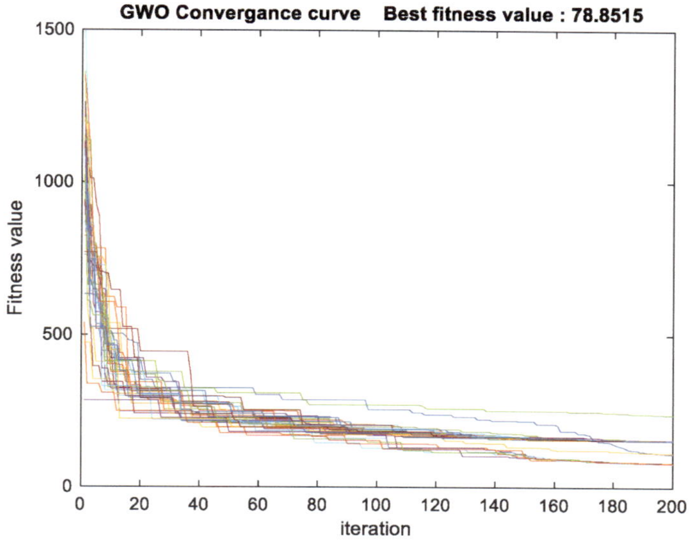

Figure 7.
GWO convergence curves for clustering IRIS data.

5. Results

Clustering is one of the unsupervised machine learning methods grouping data to the clusters. In this study, four well-known swarm-based, nature-inspired optimization algorithms have been used for clustering. In order to measure the clustering performance of the algorithms, sum of the distance values have been used. Clustering performance of the algorithms on IRIS dataset has been tested for comparison. As it has been seen in the tables, nature-inspired algorithms' solution time is not comparable with k-means. Nature-inspired algorithms are very slow because of the swarm-based run. According to the tables, PSO and GWO are faster than BBO and GA owing to the mutation and other parameters. Both PSO and GWO have fewer parameters to adapt, and they are faster and more stable than BBO and GA. In this study, no adaptation is applied for any algorithm. So if special adaptation is applied for those algorithms, the clustering performance of the algorithms will increase.

Author details

Pakize Erdogmus* and Fatih Kayaalp
Computer Engineering Department, Engineering Faculty, Duzce University, Duzce, Turkey

*Address all correspondence to: pakizeerdogmus@duzce.edu.tr

References

[1] Available from: https://www.gtac.edu.au/the-kingfisher-and-the-bullet-train-in-the-news/. Latest access

[2] Goldberg DE, Deb K. A comparative analysis of selection schemes used in genetic algorithms. Foundations of Genetic Algorithms. Vol. 1. Elsevier; 1991. pp. 69-93

[3] Kennedy J, Eberhart R. Particle Swarm Optimization. 1995. pp. 1942-1948

[4] Simon D. Biogeography-based optimization. IEEE Transactions on Evolutionary Computation. 2008;**12**(6): 702-713

[5] Mirjalili S, Mirjalili SM, Lewis A. Grey wolf optimizer. Advances in Engineering Software. 2014;**69**:46-61

[6] Aggarwal CC. An introduction to cluster analysis. Data Clustering. Chapman and Hall/CRC; 2018. pp. 1-28

[7] Dubes R, Jain AK. Clustering methodologies in exploratory data analysis. In: Advances in Computers. Vol. 19. Elsevier; 1980. pp. 113-228

[8] Twinkle G, Kumar D. Optimization of clustering problem using population based artificial bee colony algorithm: A review. International Journal. 2014;**4**(4)

[9] Ashour AS et al. A hybrid dermoscopy images segmentation approach based on neutrosophic clustering and histogram estimation. Applied Soft Computing. 2018;**69**: 426-434

[10] Milone DH et al. Clustering biological data with SOMs: On topology preservation in non-linear dimensional reduction. Expert Systems with Applications. 2013;**40**(9):3841-3845

[11] Paul R, Hoque ASML. Clustering medical data to predict the likelihood of diseases. In: 2010 Fifth International Conference on Digital Information Management (ICDIM). IEEE. 2010

[12] Breiger RL, Boorman SA, Arabie P. An algorithm for clustering relational data with applications to social network analysis and comparison with multidimensional scaling. Journal of Mathematical Psychology. 1975;**12**(3): 328-383

[13] Singh K, Shakya HK, Biswas B. Clustering of people in social network based on textual similarity. Perspectives in Science. 2016;**8**:570-573

[14] Bandyopadhyay S, Coyle EJ. An energy efficient hierarchical clustering algorithm for wireless sensor networks. In: Twenty-Second Annual Joint Conference of the IEEE Computer and Communications Societies, IEEE INFOCOM 2003 (IEEE Cat. No. 03CH37428), Vol. 3. IEEE. 2003

[15] Fahad A et al. A survey of clustering algorithms for big data: Taxonomy and empirical analysis. IEEE Transactions on Emerging Topics in Computing. 2014; **2**(3):267-279

[16] Greche L, Jazouli M, Es-Sbai N, Majda A, Zarghili A. Comparison between Euclidean and Manhattan distance measure for facial expressions classification. In: 2017 International Conference on Wireless Technologies, Embedded and Intelligent Systems (WITS), Fez. 2017. pp. 1-4

[17] Chouikhi H, Saad MF, Alimi AM. Improved fuzzy possibilistic C-means (IFPCM) algorithms using Minkowski distance. In: 2017 International Conference on Control, Automation and Diagnosis (ICCAD), Hammamet. 2017. pp. 402-405

[18] Jain AK, Narasimha Murty M, Flynn PJ. Data clustering: A review. ACM Computing Surveys (CSUR). 1999;**31**(3):264-323

[19] Eberhart R, Kennedy J. A new optimizer using particle swarm theory. In: MHS'95. Proceedings of the Sixth International Symposium on Micro Machine and Human Science. IEEE. 1995

[20] Shi Y, Eberhart R. A modified particle swarm optimizer. In: 1998 IEEE International Conference on Evolutionary Computation Proceedings. IEEE World Congress on Computational Intelligence. IEEE; May 1998. pp. 69-73. (Cat. No. 98TH8360)

[21] Clerc M, Kennedy J. The particle swarm—Explosion, stability, and convergence in a multidimensional complex space. IEEE Transactions on Evolutionary Computation. 2002;**6**(1): 58-73

[22] Zhang X, Wang D, Chen H. Improved biogeography-based optimization algorithm and its application to clustering optimization and medical image segmentation. IEEE Access. 2019;**7**:28810-28825

[23] Pal R, Saraswat M. Data clustering using enhanced biogeography-based optimization. In: 2017 Tenth International Conference on Contemporary Computing (IC3), IEEE. 2017

[24] Emary E, Zawbaa HM, Hassanien AE. Binary grey wolf optimization approaches for feature selection. Neurocomputing. 2016;**172**: 371-381

[25] Mirjalili S. How effective is the Grey wolf optimizer in training multi-layer perceptrons. Applied Intelligence. 2015; **43**(1):150-161

[26] Kumar V, Chhabra JK, Kumar D. Grey wolf algorithm-based clustering technique. Journal of Intelligent Systems. 2017;**26**(1):153-168

[27] Zhang S, Zhou Y. Grey wolf optimizer based on Powell local optimization method for clustering analysis. Discrete Dynamics in Nature and Society. 2015;**2015**(1)

[28] Fahad M et al. Grey wolf optimization based clustering algorithm for vehicular ad-hoc networks. Computers & Electrical Engineering. 2018;**70**:853-870

[29] Available from: https://archive.ics.uci.edu/ml/datasets/Iris

[30] Fisher RA. The use of multiple measurements in taxonomic problems. Annals of Eugenics. 1936;7(2):179-188

Section 2

System Design and Architecture

Chapter 2

Best Practices in Accelerating the Data Science Process in Python

Deanne Larson

Abstract

The number of data science and big data projects is growing, and current software development approaches are challenged to support and contribute to the success and frequency of these projects. Much has been researched on how data science algorithm is used and the benefits of big data, but very little has been written about what best practices can be leveraged to accelerate and effectively deliver data science and big data projects. Big data characteristics such as volume, variety, velocity, and veracity complicate these projects. The proliferation of open-source technologies available to data scientists can also complicate the landscape. With the increase in data science and big data projects, organizations are struggling to deliver successfully. This paper addresses the data science and big data project process, the gaps in the process, best practices, and how these best practices are being applied in Python, one of the common data science open-source programming languages.

Keywords: Python, big data, data science process, best practices, open source

1. Introduction

Organizations use insights derived from data to stay competitive. The big data phenomenon has made deriving insights more challenging due to the changing characteristics of the data landscape. The increased volume, variety, and velocity of big data challenge traditional information technology (IT) processes to scale and support big data analytics and data science. Big data is used by organizations as a resource in data science projects to develop new business value and insights. Big data examples include sensor data, images, text, audio, and video data. These data sources can provide new insight opportunities alone and when paired with existing data sources such as organizational data warehouses.

Data science is a competency that leverages data processing, algorithms, and math to develop insights from data. Data science is a core competency that organizations want to develop to stay competitive. While data science as a competency is growing, the practices to ensure these projects are successful have not kept up with the pace. One primary challenge is using existing software development methodologies to deliver data science projects. Applying traditional software methodologies, such as the waterfall approach, is problematic and has been identified as the one contributing factor for data science project failure; organizations are treating data science projects like other IT projects [1].

According to Saltz, current research in data science and big data has primarily focused on the use and application of algorithms and generating insights, but little to no research has occurred about tools, methodologies, or frameworks used to deliver data science projects [2]. Saltz outlined that existing tools, methodologies, and frameworks are not mature enough to effectively use in data science and big data projects [2]. This paper focuses on the changing data landscape, the data science process, and identifying best practices to accelerate the data science processing using Python. Python is an interpreter, object-oriented programming language and one of the most popular tools used in big data and data science projects.

2. The changing data landscape

The increased use of internet-connected smart devices has changed how organizations use information [3–5]. The Internet of Things (IoT) creates large amounts of data quickly from sensors embedded in devices, one of the contributing factors in creating the category of big data [5]. The rise of data science is primarily a result of big data, due to the need to analyze data, other than traditional structured data, such as text, machine-generated, and geospatial data [5]. Big data and data science go hand in hand; thus software development approaches used need to consider both [1, 4, 6]. Results of a data science project not only include insight, but also working software that needs to be deployed and supported. Analyzing the characteristics of big data highlights the challenges with traditional software development approaches.

2.1 Volume

The growth of data impacts the scope of data used in data science and software development. Scope increases project complexity where new technology is used to accommodate more data [3]. Large amounts of unstructured data cannot be easily ingested and processed using a traditional relational database for example.

2.2 Variety

Data variety becomes a concern for software development as the types of data sources to be used for development and analysis increase. The variety of data means increasingly complex forms of data such as structured and unstructured data [3–5]. Traditionally, structured data is created in rows and columns and easily understood; however, unstructured data comes in different forms, levels of details, and without clear metadata complicating the ability to understand and use [3–5].

Examples of data variety include images, IoT sensor data, clickstream, images, and event data. These data sources may be analyzed independently, but often analysis requires data to be integrated. Integrating multiple data sources with different structures increases the complexity of projects.

2.3 Velocity

The speed at which data is created is referred to as velocity. In 2014, Twitter averaged 1 billion tweets every few days [7]. Fresher data results in the ability to analyze new patterns and trends that were not possible before big data. With IoT applications, data that is 15 minutes may be too old for analysis [5]. Data acquisition becomes a challenge as traditional data acquisition focused on extract, transformation, and load (ETL) of data. Increased velocity changes the order where data is loaded first, then analyzed, otherwise known as extract, load, and then transformation (ELT) [3–5].

2.4 Veracity

Veracity refers to how accurate data is and how well the data is understood. Big data is not clearly analyzed prior to ingestion due to the volume and speed of creation, which results in data that may have credibility and reliability problems. Often metadata for big data sources do not exist. These challenges increase the complexity of deriving insights from big data sources [3–5].

Software development lifecycles have traditionally focused on requirements which drove the design, leveraging the design to develop software, testing, and then deploying the software. Projects using big data change the order in which these phases occur. Big data sources are ingested, stored, and explored first, and then requirements are determined which changes the traditional order of activities for project delivery. Using the traditional software development lifecycle for projects that include big data has failed further supporting that projects using big data need to adjust project approaches [1].

According to Mayer-Schönberger and Cukier, big data changes how the world interacts and means a disruption to what was considered normal. This disruption also means disruption to the software development processes that create the software that derives knowledge and value from data. Big data results in changes to IT processes, technologies, and people. One greater reliance observed is that data scientists need to address the complexity introduced with big data and help derive the knowledge from data [3, 4].

3. The data science process

According to Saltz, most data science processes focus on the tasks that need to be completed in data science such as the techniques to acquire and analyze data [2]. Saltz analyzed different data science approaches and found that most outlined the steps as data acquisition, cleansing, transformation, integration, modeling, analysis, and deployment [2]. The data science approaches are task-oriented, and no real evolution of the process had occurred since the cross-industry standard process for data mining (CRISP-DM) was introduced in the 1990s [2].

3.1 CRISP-DM

The most commonly used data mining process is CRISP-DM which is a process that conceptually described the stages used in data mining. Originally created to support data mining projects, it has been adapted by data scientists. There are six stages in the CRISP-DM process which are presented sequentially, but iteration is expected:

- Business understanding
- Data understanding
- Data preparation
- Modeling
- Evaluation
- Deployment

3.2 Business understanding

The start of the process, business understanding, focuses on the business value of the project. Once requirements and objectives are understood, the problem definition is created. Data science projects start with a problem to be addressed or a question to be explored. Tools to support identifying the problem include diagrams such as a decision model such as a fishbone diagram [8].

3.3 Data understanding

Once the problem to be addressed is identified, the next focus is on data source identification and collection. Data source identification includes identifying the sources, such as a transactional processing system, and the attributes needed to address the problem. Problems may involve several different data sources, which means data integration work is likely. After integration, data is profiled and statistically analyzed to determine quality, demographics, relationships between variables, and distribution. The outcome of data understanding is a definition of how the data can be used. The data understanding stage is often referred to as exploratory data analysis (EDA) [8].

3.4 Data preparation

The goal of data preparation is to create the data that is to be used in the modeling stage. Input from data understanding is used to determine the final set of attributes, often referred to as features that will be used in the model. Preparation includes integration, cleansing, and deriving of new attributes. Data preparation is iterative as the training and testing of the model may require new or changed data. The result of data preparation is the data set to be used as input into the modeling stage [8].

3.5 Modeling

As part of the modeling stage, different techniques and algorithms are used to determine the best model. Modeling goes through cycles of testing and training, where the data scientist adjusts parameters to produce the best outcomes. It may be necessary to return to data understanding and preparation stages if the model performance is not acceptable [8].

3.6 Evaluation

Model evaluation is determined based on the overall fit to the problem statement and business objectives. Evaluation is conducted by analyzing error rates, variance, and bias of the model. Often models using different techniques are compared to determine the best performing one. Once the best performing model is identified, a formal review is conducted to move to model deployment [8].

3.7 Deployment

The deployment stage is often overlooked and unplanned for but important as this is where business value is realized. Deployment focuses on a working software model that can be supported and executed on a regular frequency. Once deployed, models are monitored for performance as model accuracy will degrade because of organizational change and time. Models are often retrained once this occurs [8].

The description provided of CRISP-DM is a summary and does not highlight the granularity of effort needed for successful data science outcomes. Many organizations use CRISP-DM as a framework and add steps to each stage for software development teams to follow. There was an effort to create CRISP-DM 2.0 in 2007, but there is no new research or activity in this area [2].

4. The role of open source in data science projects

Landset et al. outlined that big data has caused IT departments to rethink how data is processed and the use of data science software [9]. Choosing the right processing framework and data science software can be challenging due to data science project requirements and that data complexity might require more than a single solution [9]. Additionally, tools available to conduct data science are many with partial functionality, limited ability to handle big data, and integrate into big data processing platforms, which contributes to the fragmentation and complexity of the data science and big data technology solutions.

Landset et al. called out that no single tool or framework covers all of the requirements of a data science project [9]. Data scientists and computer engineers need tools that support computing performance, usability, machine learning algorithm breadth, and portability. To support these needs, data scientists and computer engineers have turned to open-source tools to address the variety of requirements of data science projects. Open-source technology categories include data processing platforms and engines and machine learning tools. The Hadoop ecosystem is commonly used to address the data processing aspect of big data and data science [9]. The Hadoop ecosystem includes workflow, data collection, storage, and processing, for example. While there are many open-source machine learning tools, Python has become one of the leading programming languages used in data science as well as R and Mahout [9].

5. Challenges with big data and data science projects

To best understand what the best practices are in big data and data science projects, it is beneficial to highlight some of the challenges. Some of the challenges highlighted so far include lack of a detailed software development methodology and the characteristics of big data. Other challenges include that many data science projects are managed as a single project and the focus on reproducibility, collaboration, and communication is overlooked [10].

Andrejevic argued that big data changes how businesses and customers interact which disrupts normal business processes [11]. This disruption also means disruption to the software development processes used to create the data science models that derive data insights [11]. When organizations leverage big data, this results in changes to IT processes, technologies, and people [11]. One change observed is that data scientists are challenged to handle all activities related to the data science process including all the data wrangling activities which can consume most of the project timeline [3, 11]. Traditional IT projects normally have defined roles and activities involving several team members [3, 11].

Mayer-Schönberger and Cukier highlighted the impact of big data on organizations citing that the volume, variety, and velocity characteristics make data science and big data projects difficult to deliver using traditional IT project approaches [12]. Volume challenges how much data is ingested and consumed, variety highlights the need to support different data structures, and velocity outlines the need to ingest data as soon as it is created [12]. The need for nontraditional (non-relational databases and storage systems) data processing platforms and software that can handle big data is the reason why traditional IT processes are unsuited for data science and big data projects [12].

Some of the challenges in delivering data science and big data projects include having clear business objectives, dealing with volume, identifying what data to use, understanding opportunities to store and process data, and having clear privacy and security requirements [13]. Mousannif et al. proposed a framework for a big data project workflow that included planning, implementation, and post-implementation phases [13]. Mousannif et al. highlighted a need to focus on reproducibility of data ingestion, processing, and storage to expedite future big data projects [13].

Lowndes et al. proposed that data science and big data projects can be accelerated by focusing on reproducibility, transparency, and collaboration by implementing new processes leveraging open-source tools [10]. Lowndes et al. published the results of implementing new processes to accelerate data science for the Ocean Health Index (OHI) project which is repeated yearly to address the change in global ocean health [10]. New processes were implemented in the categories of reproducibility, communication, and collaboration and broken down further into specific tasks [10].

Lowndes et al. outlined that the following areas need to be addressed for reproducibility: data preparation, modeling, version control, and organization [10]. Data preparation includes creating and leveraging common coding routines to sort, cleanse, transform, and format data [10]. Modeling focused on standardizing to a common programming language to ensure all were using the same algorithm methods which resulted in reduced iterations on validating results [10]. Version control ensured that the team was treating the data science process as a software development process where code was tracked and change management was put in place to improve software reusability [10]. Organization was addressed through leveraging in-code documentation standards, file naming standards, and treating each data science project as a single set of common code by implementing the project function in the programming language [10].

Lowndes et al. proposed that collaboration be improved by having centralized coding repositories, common workflows for promoting code, and a centralized repository for communication [10]. Git, a widely used cloud-based version control system, was used as the common coding repository, and a Wiki was used for documenting projects [10]. Having a common project management approach was also highlighted as a benefit [10].

Lowndes et al. focused on improved team communication and effectiveness through sharing data and methods [10]. The focus was on not redoing work if the work was already completed [10]. Data sharing covered centralizing cleansed data sets for reuse and creating common data pipelines to be used for data science projects (like the OHI project) [10]. Since the OHI project is completed yearly, much of the software development work could be leveraged from the prior year in place of starting the project from scratch each time [10].

There are several gaps and lack of maturity in the data science process as outlined by the research of Landset et al., Mayer-Schönberger and Cukier, Mousannif et al., and Saltz. Based on these gaps, best practices will be proposed to accelerate the data science process leveraging Python. While other open-source programming languages could be used in the same manner as suggested in the next section, the choice of Python is not meant to recommend that Python is the only choice or best choice to use in data science projects. Python was chosen due to its popularity, performance, and portability capabilities in the data science and big data space.

6. Best practices to accelerate data science with Python

People, process, and technology are contributing factors to data science complexity. Data scientists are expected to multiskilled resources knowing statistics, big data platforms, pipeline development, and deep learning neural networks. The primary process leveraged for data science is CRISP-DM which has not really changed since it was introduced in the 1990s. Lastly, there is so much available technology to use in data science it boggles the mind. While these problems will take time to solve, there are some best practices that can be leveraged to make data science less complex and more scalable in organizations. Best practices will be addressed in general as well as with Python.

6.1 Team collaboration

Data science initiatives tend to be done as independent efforts or one-off projects. Data scientists often work as the project manager, data wrangler, software developer, data engineer, tester, and do the data science as well. Data scientists cannot be effective assuming all these roles. With data wrangling (cleansing, transformation, formatting, etc.) taking up more than half the project, the data engineer has emerged as a key role in supporting the data science process. Assign a data engineer to handle the data wrangling, and let the data scientist focus on data science. Additionally, data science initiatives are projects. Ensure the data science initiative has a lead who can remove barriers, deal with stakeholders and get the required subject matter experts to support the data science project. Collaboration is key in making progress and valuable data science results [10, 14].

6.2 Why Python?

Python is an interpreter, object-oriented programming language introduced in 1991 and has emerged as one of the leading open-source data science tools used by data scientists [15]. Python has become a leading data science tools for several reasons. As an open-source tool, Python is freely available and modifiable, keeping costs low and promoting rich features through the open-source community [6].

Python is easy to learn. Although it is a programming language, the syntax is easy to adopt, especially by programmers, and through studies, the learning cycle tends to be shorter than a comparable data science tool—R. R is another open-source programming environment specializing in statistical computing and graphics. Both tools are comparable in many areas; however, Python has emerged as being more scalable and portable [15].

Scalability and portability are important in the data science community. With big data characteristics such as volume, velocity, and variety, data science tools need to be robust. Scalability refers to the ability to handle growing computing requirements, and portability is the ability to easily run on multiple computing platforms. Python has libraries and packages that support fast computing, and it runs on the common operation systems such as Linux, Windows, Unix, and macOS [15]. While Python is discussed as a data science tool, it is also a programming language used in the development of applications.

Team collaboration was mentioned as a best practice in data science; however, collaboration is inherent within open-source communities. Python has a deep reach in the data science community with data scientists contributing to creating new libraries and code routines. In addition, the tech companies often choose Python to release new functionality first. Google, for example, released its deep learning package TensorFlow first in Python, setting a trend [14, 15]. The Python community provides an avenue for new data scientist or even seasoned ones to find solutions to data science problems. Communities often exchange questions and answers on websites such as Stack Overflow or share code on public repositories such as Git [14, 15].

Through the data science process, data scientists must visualize data relationships, and Python also supports robust visualization options [14]. Libraries exist to support multiple graphing options such as charts, graphs, and interactive plots [15]. Libraries are collections of methods and functions that data scientists can leverage without having to write new code.

The most significant factor supporting Python as the leading open-source data science tool is the number of libraries available for data scientists. These libraries, as mentioned, provide prepackaged methods and functions, and several libraries available in Python have enabled data scientists to leverage the latest machine learning algorithms (such as TensorFlow), manipulate data easily, and create data science models that perform and scale [15].

While there are several reasons why Python is popular with data scientists, other tools such as R, Scala, and Ruby are available that provide similar functionality as Python. The intent of the following section is to show how functionality in tools can accelerate the data science process. The scope of this research is not to compare Python to other languages, but to illustrate how to accelerate the process of data science.

6.3 Libraries in Python

Python has a plethora of libraries available for use, but there are a few that can accelerate the data science process and have become the set of tools that data scientists leverage. Libraries that are heavily used by the data science community include NumPy, SciPy, pandas, Matplotlib, and Scikit-learn. The "Py" at the end of Python libraries is pronounced "Pie" [15].

NumPy is a library created by Travis Oliphant that is leveraged by most of the other data science libraries. NumPy provides a large set of mathematical functions that operate on multidimensional array data structures. Arrays allow data to be stored in blocks across multiple dimensions which enable mathematical operations to be applied to matrices and vectors. Arrays support vectorization which allows fast math operations supporting scalability and performance which is valuable in solving data science problems [15].

SciPy is another library created by Oliphant, Peterson, and Jones. SciPy leverages NumPy functionality and includes additional algorithms, matrix processing, and image processing; SciPy and NumPy are often paired together in data science where NumPy provides enhanced performance and SciPy an extended library of mathematical functions and advanced processing on data types [15].

Pandas focuses on object data structures. Most working in data management default to thinking of data in rows and columns. Pandas handles processing of data tables with different data types (very similar to a relational database table) as well as time series. Pandas enables easy data manipulation without the constraints of a relational database. Slicing, dicing, dropping, and adding elements, reshaping, joining, and aggregating data is much easier, leveraging the object data structures called dataframes. Dataframes are commonly used in R as well [15].

Matplotlib is another library that leverages NumPy. Matplotlib is highly used for charting, graphing, and time series analysis, especially in the data understanding and preparation phases of the data science process [15].

Scikit-learn (also referred to as Sklearn) contains the core data science method and functions for Python [15]. Scikit-learn contains methods and functions that cover supervised and unsupervised algorithms, model selection, model validation, and final evaluation of performance. Several modules exist that data scientists should be aware of such as preprocessing and feature extraction that contribute to accelerating the data science project. Specific examples will be addressed as part of the data science process. The stages of CRISP-DM will be used to address when best practices get leveraged. Although CRISP-DM starts with business understanding, the application of Python typically starts in data understanding.

6.4 Data understanding

Data understanding focuses on the data collection and analysis of the data sources to be used in the data science project. In data understanding, the three areas that can help accelerate the data science process are data profiling, data visualization, and data preprocessing. Data profiling is the process of gathering statistics and demographics about data sets. Profiling provides the ability to assess data quality and understand how attributes are populated. Library pandas_profile can be leveraged to complete data profiling. Data profiling includes data set info, variable types, descriptive statistics, and correlations between numeric variables [15]. An example of the output is listed in **Figures 1** and **2**.

6.5 Data preparation

Data preprocessing focuses on scaling, normalization, binarization, and one hot encoding. Scaling is applied when the values of potential features have a large variance between random variables. Data normalization is used to adjust the values in a feature vector so that they can be measured on a common scale. Binarization is used to convert a numerical feature vector into a binary vector such as true or false. One hot encoding is a process by which categorical variables are converted into a form that could be provided enables better prediction by algorithms. One hot encoding determines feature frequency, identifies the total number of distinct values, and then uses a one-of-k scheme to encode the values [15, 16]. Preprocessing data accelerates the data science process by minimizing the number of iterations in the modeling stage. The Scikit-learn library has methods for scaling, normalization, binarization, and one hot encoding such as sklearn.preprocessing.

Automating feature selection is the process of reducing the number of input variables used by predictive models. Feature selection can be a time-consuming process for data scientist. Automatically selecting those features that are most useful or most relevant for use in the analytical problem accelerates the data science process. Scikit-learn has multiple methods that support overfitting, improve accuracy, and reduce the training time of models such as sklearn.feature_extraction [15, 16].

Dataset info

Number of variables	19
Number of observations	94682
Total Missing (%)	0.0%
Total size in memory	13.7 MiB
Average record size in memory	152.0 B

Variables types

Numeric	8
Categorical	2
Boolean	7
Date	0
Text (Unique)	0
Rejected	2
Unsupported	0

Figure 1.
The data set information and variables types can be created using the pandas_profile library.

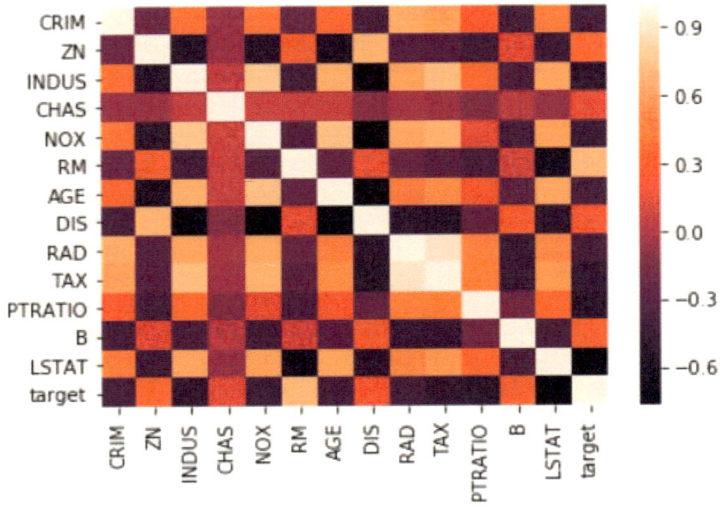

Figure 2.
Data visualization using the Matplotlib and Seaborn libraries is highly effective in the data understanding stage. Both libraries support multiple charts and graphs to visualize data relationships. Using the Matplotlib library, a correlation heat map can be created to demonstrate correlations within a data set.

6.6 Data pipelines

When data science is done as an independent effort, much of the work in the data understanding and preparation phase of a data science project is done without the end-state vision. Data understanding and preparation mimics many of the development activities that are used to develop data pipelines. The activities to develop the data pipeline include acquiring the data, exploring the data for deep understanding and value determining, integration, transformation and formatting to get the data ready to be consumed by the model. Why not develop the pipeline as part of the project? At the end of the data science project, the expectation is a working data science model. The model will have no value if the data pipeline to produce the input is not available and ready for production deployment.

With data understanding and preparation taking more than 50% of the project timeline, data scientists need to have access to the correct data in the correct format. Not only will developing the pipeline as part of the project prepare the data for production consumption, but it may also be leveraged for other data science projects.

Another area to review is existing ETL or pipelines in the data warehouse environment. Data science leverages all kinds of data, and often models use existing transactional data enriched with big data sources. Reusing existing ETL or pipelines promotes consistency and reduces the development work in the data science project. Pipelines in Python are an easy way to automate common and repeatable steps such as the preprocessing steps. One pipeline library that can be leveraged in Python is Luigi.

One of the challenges with data pipelines is pipelines contain components that need to be linked together for processing. Many of the components in a pipeline support long-running job, streaming of data, and running machine algorithms that may fail. Luigi, the pipeline library, can help link many of these pipeline components together or provide the workflow management so that the components can be run and managed as a single pipeline. Workflow management handles the dependencies between the components. As part of the Luigi library, templates are provided that provide support for long-running jobs in Python. Luigi also contains file system abstractions for the Hadoop File System (HDFS) which ensures the pipeline will not fail holding incomplete data. Luigi also includes a Visualizer page that provides a visual status of the pipeline and provides a visual graph of the pipeline [15].

6.7 Modeling and evaluation

Scikit-learn is the library that is leveraged for modeling and evaluation. The data scientist will choose the modeling approach or algorithm to be used for the data science problem; however, tuning the model and choosing the best performing model can be a challenge. Scikit-learn has several functions and methods that can be leveraged to expedite this stage of the data science process. One example is train_test_split which supports the random creation of training and test files [15].

Training and test files are used to "fit" the model, and through this process different settings or hyperparameters are tuned to improve the accuracy of the model. This causes overfitting where performance is no longer generalized in the model. The approach to avoid overfitting is called cross validation. A function in Scikit-learn can be leveraged called cross_validate where another data is held out, referred to as the validation data set. The cross_validate function can be used where the training is completed, but evaluation is completed on the validation set. Once appropriate accuracy is achieved, the test set is used for final evaluation [15].

Model and feature selection can also be time-consuming activities in the data science process. Scikit-learn can also be leveraged to evaluate an algorithm's performance as well as to select the best fit parameters using cross validation. Another method that can be used is GridSearchCV. GridSearchCV is used to wrap an estimator, where it selects parameters from training file to maximize the score; GridSearchCV can be used as part of a pipeline to combine several transform components and estimators to create a new estimator as well. Scikit-learn offers many options for data processing, model selection, and model validation. It also includes a complete set of methods to support many different modeling algorithms [15].

6.8 Deployment

Software engineering principles should be applied to both the data pipeline and the data science model. The likelihood of using big data where volume, velocity, and variety need to be addressed means that the data science project scope is not only to produce working software, but also software that has quality and can scale.

The characteristics of data science software quality specifically focus on supportability, reusability, reliability, portability, and scalability. Supportability focuses on the ability of IT operations to address maintenance and failure issues. Reusability is the ease with which software can be reused other data science projects. Reliability is the frequency and criticality of software failure, where failure is an unacceptable behavior occurring under permissible operating conditions. Portability is the ease with which software can be used on computer configurations other than its current one. Last, scalability is the ability to handle performance demand without having to re-architect the software. All these characteristics should be applied to data science pipelines and models [10].

Scalability tends to be the largest challenge with machine learning algorithms. The expectation with machine learning algorithms is that as data increases, the algorithm addresses the volumes in an efficient way where the run time increases linearly. Machine learning algorithms that do not scale increase running time exponentially or simply stop running. One way to address the scaling problem is to use out-of-core learning [15].

Out-of-core learning uses a set of algorithms where data that does not fit into memory can be stored in another location such as a repository or disk. Scikit-learn includes functionality that supports the streaming of data from storage and iterative learning [15].

6.9 Summary of Python libraries to accelerate data science

As mentioned prior, there are many tools available to data scientists, and the landscape is evolving daily [9]. For this research, Python was chosen due to popularity, and other open-source languages may have similar capabilities. Several different Python libraries were recommended that could accelerate the data science process. A summary of these libraries follows.

There are five core libraries available in Python that accelerate performance in the data science process. NumPy provides a large set of mathematical functions that operate on multidimensional array data structures. SciPy, which is often paired with NumPy, includes additional algorithms, matrix processing, and image processing functionality. Pandas enables the use of object data structures in rows and columns. Matplotlib is a visualization library which pairs well with NumPy. Scikit-learn contains many different machine learning algorithms.

While most of these core libraries will be used in the phases of data science, there are several packages in each library that help accelerate some of the challenges with the data science process. In the data understanding stage, both Matplotlib and Seaborn support many different types of visualizations. Pandas_profile, part of the Pandas library, quickly completed the profiling process exposing data set demographics.

Data preparation tends to make time which could be better spent on data science modeling. Leveraging the packages of preprocessing and feature_extraction as part of Scikit-learn helps reduce some of the work. Preprocessing quickly handles scaling, normalization, and binarization of variables. Feature_extraction evaluates different features for value to reduce training time. Focusing on building a reusable pipeline in the data preparation stage can accelerate the deployment stage. Luigi is a pipeline package that can automate code modules, create workflow, and help with support of data pipelines once in production.

As part of modeling and evaluation, packages from Scikit-learn are used to automate training and testing data set creation using train_test_split. Both the cross-validate and GridSearchCV can be used to evaluate models and compare models to get the final recommendation. Once the model is ready for deployment, the use of out-of-core learning can be used to maximize the use of computing resources in production as data science models tend to heavily use computing power.

7. Conclusion

Current research in data science and big data has primarily focused on the use and application of algorithms and generating insights, but little to no research has occurred about tools, methodologies, or frameworks used to deliver data science projects. Analyzing the characteristics (volume, variety, and velocity) of big data highlights the challenges with traditional software development approaches. Results of a data science project not only include insight, but also working software that needs to be deployed and supported; thus software engineering practices should not be avoided. Leveraging tools such as Python can help accelerate the data science process. Python has become a leading data science tool for several reasons, primarily due to the functionality that is created quickly to address the data science community needs.

DOI: http://dx.doi.org/10.5772/intechopen.84784

Author details

Deanne Larson
Larson and Associates, LLC, Bothell, WA, USA

*Address all correspondence to: larsonelink@aol.com

References

[1] Demirkan H, Dal B. The Data Economy: Why Do So Many Analytics Projects Fail? 2014. Retrieved from http://analytics-magazine.org/the-data-economy-why-do-so-many-analytics-projects-fail/

[2] Saltz JS. The need for new processes methodologies and tools to support Big Data teams and improve Big Data project effectiveness. 2015 IEEE Santa Clara, CA, USA: International Conf. on Big Data; 2015

[3] Abbasi A, Suprateek S, Chiang R. Big data research in information systems: Toward an inclusive research agenda. Journal of the Association for Information Systems. 2016;**17**(2):i-xxxii

[4] Davenport TH. Analytics 3.0. Harvard Business Review. 2013;**91**(12):64-72

[5] Halper F. Next-Generation Analytics and Platforms for Business Success. TDWI Research Report. (2015). Available from: www.tdwi.org

[6] Gartner Research (2015). Gartner Says Business Intelligence and Analytics Leaders Must Focus on Mindsets and Culture to Kick Start Advanced Analytics. Gartner Business Intelligence & Analytics Summit. 2015

[7] Abbasi A, Adjeroh D. Social media analytics for smart health. IEEE Intelligent Systems. 2014;**29**(2):60-64

[8] Marbán O, Mariscal G, Segovia J. A data mining & knowledge discovery process model. In: Data Mining and Knowledge Discovery in Real Life Applications. Vienna, Austria: I-Tech; 2009. pp. 438-453

[9] Landset S, Khoshgoftaar TM, Richter AN, Hasanin T. A survey of open source tools for machine learning with big data in the hadoop ecosystem. Big Data. 2015;**2**:2-24. DOI: 10.1186/s40537-015-0032-1

[10] Lowndes JSS, Best BD, Scarborough C, Afflerbach JC, Frazier MR, O'Hara CC, et al. Our path to better science in less time using open data science tools. Nature Ecology & Evolution. 2017:**1**(6);0160-0167. DOI: 10.1038/s41559-017-0160

[11] Andrejevic M. Big data, big questions the big data divide. International Journal of Communication. 2014;**8**:17. Available from: https://ijoc.org/index.php/ijoc/article/view/2161/1163

[12] Mayer-Schönberger V, Cukier K. Big Data: A Revolution that Will Transform How we Live, Work, and Think. Boston, MA and New York, NY: Eamon Dolan/Houghton Mifflin Harcourt; 2013

[13] Mousannif H, Sabah H, Douiji Y, Younes OS. Big data projects: Just jump right in! International Journal of Pervasive Computing and Communications. 2016;**12**(2):260-288. DOI: 10.1108/IJPCC-04-2016-0023

[14] Jagadish H, Gehrke J, Labrinidis A, Papakonstantinou Y, Patel J, Ramakrishnan R, et al. Big data and its technical challenges. Communications of the ACM. 2014;**57**(7):86-94

[15] Joshi P. Python: Real World Machine Learning (Kindle Locations 8418-8423). Birmingham, UK: Packt Publishing; Kindle Edition

[16] Pedregosa F, Varoquaux G, Gramfort A, Michel V, Thirion B¨, Grisel O, et al. Scikit-learn: Machine learning in python. Journal of Machine Learning Research. 2011;**12**:2825-2830

Chapter 3

Software Design for Success

Laura M. Castro

Abstract

Technical books focus most of the times in technical stuff, as one should expect. However, this creates the illusion that technology is somewhat free of bias, always neutral, thus fitting everyone. Reality, later on, when the product is already there, proofs us otherwise. Inclusion and representation are crucial from the design and modeling stages. Visibility of minorities or underrepresented groups is on the rise, yet so much of the way is left for us technicians to walk. In this chapter, we will analyze, from an architectural point of view, which non-functional requirements are most sensible to this and how to start the conversation about them to maximize the possibilities for success of our software products.

Keywords: inclusion, visibility, representation, software architecture, non-functional requirements

1. Introduction

Software is omnipresent. From personal computers and laptops, it has extended its presence to tablets, smartphones, smart watches, and wristbands. From software packages delivered on CDs, it has moved to apps and services which run uninterruptedly on remote servers. From our professional workplace, it has conquered our personal lives, relationships, and leisure activities.

We could see this as proof of the success of the software industry, the technology revolution. But is it? What does success represent exactly, in societal terms? Are we solving people's problems, or rather are we creating new ones? In order to make this argument more objective, we need to define success. We can argue that success of the software industry is proved by its constant innovation. But innovation is not necessarily equal to progress, which should be the key indicator in terms of societal benefit.

According to Wikipedia, "progress is the movement towards a refined, improved, or otherwise desired state (...), the idea that advancements in technology, science, and social organization can result in an improved human condition" [1]. Arguably, we seem to be constantly producing advancements in technology, but it has also become evident that the technology advancements we are producing are not improving the condition of all humans equally. Mass media is regularly hit by news where "algorithms" are revealed as biased, showing behaviors which are sexist, racist, LGBTI-phobic, etc. Software leaves minorities out, and apps discriminate on bases of age or socioeconomic status [2]. Be it in recruiting [3], evaluating risk for a financial product, assigning probability of crime involvement [4, 5], targeting adds [3], or classifying our pictures [6], the direct consequence of these errors is blatant failure.

IntechOpen

Of course, there are many angles to this systemic problem. In this chapter, we will analyze how we can contribute to progress, ensuring the success of our software product, from the perspective of software architecture.

Software architecture is the part of software development which defines the high-level decomposition of a system into a set of functional components and describes its responsibilities and interactions. A software architect is thus responsible for selecting those components, defining said interactions, and describing the constraints that operate over both of them. These decisions are grounded on both functional and non-functional requirements of the software and will serve as basis for the design, implementation, and testing stages (no matter which development cycle is used).

Consequently, one of the most important skills of a software architect is asking the relevant questions which answers can make the difference between product success and failure. The said questions need to provide confidence in that both functional and non-functional requirements are correctly elicited, understood, and quantified, as a mandatory previous step to allow their correct development and validation. In particular, failure to properly define non-functional requirements (also referred to as "system requirements") is the third cause of software project failure [7].

In the remainder of this chapter we will go over the definition of non-functional requirement and provide a taxonomy for practical use. We will identify the non-functional requirements that are more closely related to software success in terms of societal progress. We will discuss them and provide insights on how to extract and test them.

The aim we pursue by doing this is to hand in a handbook of rules, an enhanced checklist, that would be useful for practitioners and future professionals that want to specialize in the area of software architecture. We trust the contents that follow will spark their interest in and concern about really successful software, as well as be a useful guide in building it. However, by keeping our technical level purposely abstract, we aim to make this chapter readable for the general public as well, a general public who, as massive consumer of software products, can also benefit from awareness about what kind of successful products they can and should be demanding from the software industry.

2. Software architecture: it's all about non-functional requirements

Software requirements, as defined by the IEEE [8], are "a condition or capability that must be met or possessed by a system or system component to satisfy a contract, standard, specification, or other formally imposed document." As we can see, no explicit mention is made as of the *character* of the said condition or capability. Traditionally, the software industry showed a tendency to focus on functional requirements, which are defined as combinations of behaviors between inputs and outputs [9]. Requirement elicitation and other software development and software life cycle management practices provided support for functional requirements in terms of formalisms like use cases, user stories, etc. The focus was on *what* software had to do, rather on *how* it was supposed to do it.

It is the definition of software architecture as a critical part of the software development process [10] which brings attention to non-functional requirements. Non-functional requirements (sometimes referred to as "quality requirements") are defined as criteria to be used to judge the operation of the system, rather than specific behaviors. It is this system-wide relevance that makes most non-functional requirements *architecturally significant* [11], since they impose constraints on the design or implementation. **Figure 1** shows a common taxonomy of non-functional requirements.

However, we are still failing to systematically build a software that is a success, at least in the terms we were discussing in the previous section. Both what things our systems do and how they do them reveal those omissions, biases, and plain discrimination we would very much like to eradicate.

From the software architecture perspective, this is because **Figure 1** shows a very narrow view of non-functional requirements. Compare this with **Figure 2**, which presents a more exhaustive relation of parameters of interest for any software application.

In the following subsections, we will traverse the taxonomy in **Figure 2** that extends that of **Figure 1** beyond the shadowed area, to provide insights on what might be missing from our products if we overlook them.

2.1 Revisiting product requirements

The fact that product-focused non-functional requirements have received more attention than the rest of the extended taxonomy of non-functional requirements shown in **Figure 2** does not mean they cannot and should not be revisited under the "societal success" mindset. We now see how.

Usability. In software engineering, usability is meant to quantify the quality of user experience and as such is typically described in terms of effectiveness (i.e., number of goals that can be achieved using the software) and efficiency (i.e., time required to complete said goals), alongside with other less objective parameters such as

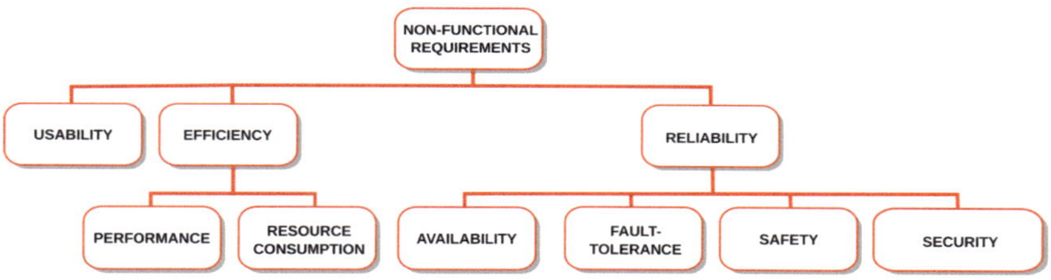

Figure 1.
Traditional taxonomy of non-functional requirements.

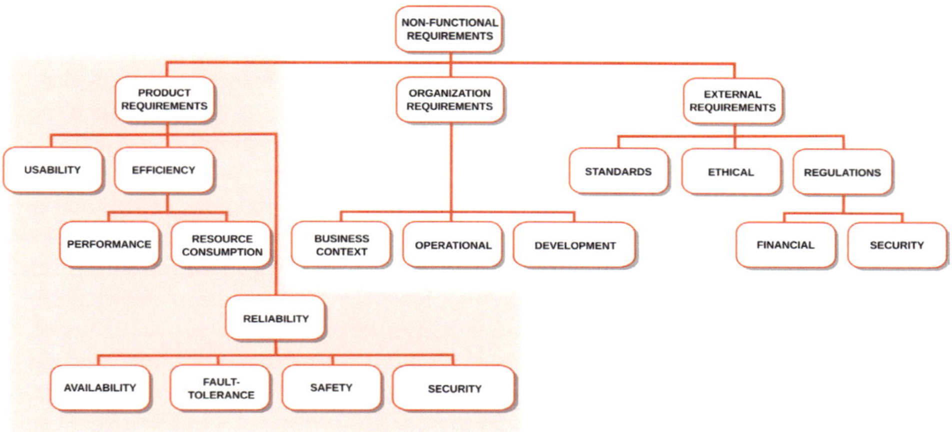

Figure 2.
Extended taxonomy of non-functional requirements.

satisfaction or perceived elegance. Usability should, however, be measured for different types of users, instead of a "regular user" or "normal user," which is often the case. The recommendation here is to look beyond our idea of typical user and consider the widest user base possible, featuring users with different levels of acquaintance with technology, from different age ranges, and health conditions (including transitory states such as pregnancy) and identifying with different gender options, with different socioeconomic backgrounds, and different sensor, motor, and mental skills.

Performance. Although different meanings can be associated with the term, the most common one is that of software responsiveness and is usually quantified in terms of requests or actions per unit of time. There are different and well-known approaches to controlling performance: from resource demand management (i.e., adjusting the ratios at which a component or subsystem generates events or polls or the context for information, explicitly limiting execution time, etc.) and resource arbitration (i.e., fixed or dynamic priorities) to its effective management (i.e., exploiting concurrency, duplicating data, or processes). Performance should, however, also take into account the consequences of these well-known strategies beyond the improvement or preservation of certain response times. When such response times have a human user at the other end, all the same we ought to take into account their expectations and perceptions of responsiveness. In this case, these can vary widely, providing further chances for improvement by alleviating the performance demands in some cases, which can be advantageously used to tend to other user profiles. Additionally, the consequences of performance degradation should also be contemplated under the light of the people that would suffer them, since depending on the kind of service we are providing and to whom, it might be more or less critical to comply.

Resource consumption. Similar to performance, resource consumption (be it computing power, volatile or nonvolatile storage, network access, bandwidth, etc.) has two sides to it. Whenever our system or software product needs to preserve resources and we aim to optimize them, unexpected consequences on the widest possible use base should be scrutinized. It is more likely that the problem is the reverse, since it is when we increase the demand for resources (or simply do not limit them) that we are more likely to implicitly exclude sectors of the population which might not have access to them. But in the context of excess of energy consumption that surrounds us, to which the prevalence of technology is no foreign actor (rather the contrary), every system and software should be as energy-aware as possible. Treating resources as unlimited is never a good idea, and it is not socially responsible either.

Availability. The most commonly understood definition of availability in the context of software engineering is the proportion of time a system or application is in a functioning condition, that is to say, capable of providing to its users the answer or services it is meant to, within acceptable conditions (i.e., usability, performance). The definition of this non-functional requirement makes sense considering that error-free software is virtually nonexistent. When we assume there will be errors, we need to define to which extent the presence of this errors can or will affect the normal operation of the system. For some systems, it is okay to be down for a few minutes, hours, or even days (if it is, e.g., software used in a factory which is down during the weekend). How flexible we can be about availability with regard to our software product depends on both the expectations of the users about it and the consequences of violating those expectations. In some cases, not being available might mean the users will turn to use our competitors' product instead (with the risk of not coming back); in others, it might affect their lives, threaten their security, or putting them on harm's way. Once more, we need to consider the broad population when leveraging said expectations and not only our "normal user."

Fault tolerance. Closely related to availability, fault tolerance is the capability of operating properly in the event of (internal) failure(s). The term usually helps us to stress that availability is not a black vs. white kind of situation, since the operating quality (i.e., performance, resource consumption) of a system might decrease proportionally to the severity of the failure(s). When we design for success with fault tolerance in mind, we aim to avoid that no undiscovered error in the software should be able to cause a total breakdown. With regard to societal success, the same considerations as with general availability apply.

Safety. Of course we never mean for our software to pose a risk to its users nor actively nor as a consequence of malfunctioning or unavailability, but actively considering this possibility during the whole development process involves considering safety as one of its requirements. Formally, however, safety includes not only not harming (no matter how severely) people but also goods and/or the environment [12, 13]. In a way, this links back to resource consumption in the energy-awareness aspect that we mentioned before. We could even argue whether introducing new technology where it does not bring societal progress, just for the sake of it, is not *safe*, since the environmental impact of the volume of technology we consume is already too high [14, 15]. Better approaches would always involve reusing or repurposing already existing technology, which is also less likely to exclude less privileged groups of population.

Security. Admittedly one of the major challenges in software engineering nowadays, we can informally define security as the resistance of a system or application to unauthorized uses, while operation is still granted for legitimate ones. There are several aspects of this claim that might be jeopardized if the diversity of the population is not properly accounted for, the most important of which would be to wrongly classify a legitimate request for an unauthorized one [16, 17].

2.2 Refocusing organization requirements

In the previous subsection we have gone over the "classical" non-functional requirements that we have more specifically labeled as product non-functional requirements. There are two more categories of non-functional requirements to consider, one of them being those non-functional requirements derived from our organization. We discuss them next.

Business context. Whether our organization is a startup, fast-growing spin-off, an enterprise with sustained trajectory, or a business in trouble has a big influence on the goals, value, and time-to-market elements that define success in traditional terms. From this lens, a critical look is also needed, in order to detect room for improvement within. In this case, we think in terms of the people who form our teams, their profiles, and the roles they play. And, different from what we have seen in the previous section, we aim not only to reflect societal diversity, but we should strive to improve it in terms of equality and representation. Diverse teams and people from unrepresented groups in positions of decision-making and power can be our most strategic business advantage.

Operational. How and why we organize the internal functioning of our organization will have an impact in our products and their quality. Identifying the essential capabilities (or lack thereof), performance measures, and actions to be taken for improvement cannot be done without explicitly accounting for diversity and work-life balance, which in turn are key to workplace satisfaction and commitment. If so, we risk coding into the so-called "company culture" a set of barriers for employees that "are not the norm." But why, if we have agreed there is not one "normal user," should we assume there is one "normal employee"? The answer is easy: we should definitely not.

Development. Among the different product development methodologies and life cycles that have been defined in software engineering, there is arguably no silver bullet. It is more a matter of finding the most suitable match between product requirements, business needs, and operational structure. When referring to the literature on software development, this is often remarked as being the case, but then we straightaway proceed to talk about iterations, sprints, stakeholders, minimum viable products, etc. without ever relating these concepts to the composition of our development teams. Software products are made by people and same as shoes or clothing, hardly ever one size fits all. In other words, the best software development approach will be that in which all of our diverse (see operational requirements) development team can be most productive at.

2.3 Advancing external requirements

In this last section of this chapter, we turn to external factors. We do this because none of us technology makers live in the vacuum, and our actions are subjected to societal norms and laws and in turn influence how societal norms and laws evolve. This means we share a double responsibility: on the one hand, diligently complying with the former, and on the other hand, challenging the status quo when it is necessary and advancing it.

Standards. Whether they dictate norms, conventions or requirements for data format, storage or exchange, or for service provision, interfacing, or requirements, standards play a fundamental role in software interoperation, especially if they are internationally recognized and publicly available (i.e., open). As software creators, we are responsible not only for being aware of which standards affect the areas we are deploying our software on and/or the activities it provides support to. What is more, implication of software agents of all sizes, small included, in standardization processes, is very much needed in a world in which, more often than not, conventions are imposed by big players or agreed upon among few of them behind closed doors. Paradoxically, non-functional product requirements are not usually enforced for standards themselves, which have a reputation for lacking usability, for instance. When referring to executable or interactive elements, standards or standardization efforts should always be accompanied by software tests. Implementing tests for standard specifications is a way of disambiguating them and stress-testing them. And of course, whenever a standard features or refers to any aspect of what a persona is, the assumptions that may be underlying need to be contrasted against the widest definition possible.

Ethical. When we discussed safety, we mentioned that introducing new technology just for the sake of it could not be considered *safe*. Hence, nor can it be considered *ethical*, we add now. More and more it is the case that CS studies feature courses that introduce future professionals to the concepts of professional ethics in the context of technology and software. However, many universities and academic institutions still do not offer them, and it is not reasonable to assume that every professional involved in technology and software creation has or will have a university background. The lack of a universal ethics, so to speak, is likely to make this the most subjective and controversial non-functional requirement of all. However, it is undeniable that we cannot look at software as a mere tool anymore, but rather a piece of technology that embodies the ethical commitments of those who make it and those who decide it should be made and used. The ethical aspects surrounding software products have two aspects to them: (a) whether the development of the product itself is contextually right or wrong or (b) whether the development of the product significantly affects the life or balance of power of or between individuals. Focusing on the latter case, we here advocate once more for a holistic approach to what individuals we bear in mind when analyzing the issue. As for the former, there are two main perspectives: (a) use software in particular, and technology in general, to eliminate or reduce suffering and maximize well-being and happiness for the greatest number of people and (b) use software in particular, and technology in general, to follow society's universal rules. The first, however, poses a great opportunity that the second misses: the chance to challenge the status quo and advance society's rules themselves, by pursuing the common and greater good. Paraphrasing a renowned Sci-Fi series, "the need of the many outweigh the need of the few," but the need of the few cannot be consistently overlooked: that's the way minorities are forever discriminated. Our non-functional ethic requirements need to address whether our software is *right*, but also whether it is *just* and *fair*.

Financial regulations. The prevalence of technology, when applied to software that runs as a service, means that we might be providing services to an international use base before we actually give some thoughts to the financial implications of this in terms of tax declarations, client rights, anti-monopoly legislation, etc. These are much specific issues than those of ethical non-functional requirements but still need us to explicitly identify and decide how to take care of them, from the different possibilities that we might have before us.

Security regulations. In line with the financial non-functional requirements, user-privacy and user-data preservation laws might affect our software products regardless of whether we operate beyond the scope of a single country or not. Furthermore, cybersecurity and privacy awareness is on the rise, so the context and actual rules we might need to oblige to or enforce are subjected to far greater dynamism that those of financial nature. This needs to be considered in terms of maintenance and product life.

3. Conclusion

Technology is not neutral. A biased development team, organization, societal context, etc. will most likely produce biased software. Biased technology perpetuates damaging stereotypes, hinders the empowerment of minorities and underrepresented groups, and ultimately delays innovation and progress. This can hardly be considered successful [18]. So far, the most effective ways of fighting biases in technology that we know are (a) being aware of said biases, in all their shapes and forms, and (b) striving to have as much diverse development teams and organizations. However, future possibilities may include the perspective of automated testing of fairness [19].

From the perspective of software architecture, one of the critical stages in software development, we can work toward the construction of less biased software by carefully analyzing the non-functional requirements that are relevant to our product. By first extending the traditional taxonomy of non-functional requirements (much focused on product requirements alone) and then (re)visiting it one by one, we have shed some light on how a software architect can contribute in this very important endeavor. We have provided a sort of exhaustive but high-level checklist that (a) practitioners and future professionals can use when analyzing and designing their systems and applications and (b) users can use to empower themselves in claiming that the whole software industry evolves to a higher level of responsibility toward not only innovation but progress.

Author details

Laura M. Castro
Universidade da Coruña, A Coruña, Spain

*Address all correspondence to: lcastro@udc.es

References

[1] Progress [Internet]. En.wikipedia.org. 2019. Available from: https://en.wikipedia.org/wiki/Progress [Accessed: 12 April 2019]

[2] Amazon scraps secret AI recruiting tool that showed bias against women [Internet]. U.S. 2019. Available from: https://www.reuters.com/article/us-amazon-com-jobs-automation-insight/amazon-scraps-secret-ai-recruiting-tool-that-showed-bias-against-women-idUSKCN1MK08G [Accessed: 12 April 2019]

[3] UK government probes algorithm bias in crime, recruitment, and finance [Internet]. Uk.finance.yahoo.com. 2019. Available from: https://uk.finance.yahoo.com/news/uk-government-probes-algorithm-bias-crime-recruitment-finance-000153694.html [Accessed: 12 April 2019]

[4] Google Photos labeled black people 'gorillas' [Internet]. Eu.usatoday.com. 2019. Available from: https://eu.usatoday.com/story/tech/2015/07/01/google-apologizes-after-photos-identify-black-people-as-gorillas/29567465/ [Accessed: 12 April 2019]

[5] Simonite T. Study Suggests Google's Ad-Targeting System May Discriminate [Internet]. MIT Technology Review. 2019. Available from: https://www.technologyreview.com/s/539021/probing-the-dark-side-of-googles-ad-targeting-system/ [Accessed: 12 April 2019]

[6] 5 Apps That Have Rampant Discrimination Built In [Internet]. Cracked.com. 2019. Available from: https://www.cracked.com/blog/5-apps-that-have-rampant-discrimination-built-in/ [Accessed: 12 April 2019]

[7] Why Software Fails [Internet]. IEEE Spectrum: Technology, Engineering, and Science News. 2019. Available from: https://spectrum.ieee.org/computing/software/why-software-fails [Accessed: 12 April 2019]

[8] IEEE 610.12-1990 IEEE Standard Glossary of Software Engineering Terminology [Internet]. Standards.ieee.org. 2019. Available from: https://standards.ieee.org/standard/610_12-1990.html [Accessed: 12 April 2019]

[9] Fulton R, Vandermolen R. Airborne Electronic Hardware Design Assurance. Oakville, Ontario (Canada): CRC Press; 2017

[10] Perry D, Wolf A. Foundations for the study of software architecture. ACM SIGSOFT Software Engineering Notes. 1992;**17**(4):40-52

[11] Chen L, Ali Babar M, Nuseibeh B. Characterizing architecturally significant requirements. IEEE Software. 2013;**30**(2):38-45

[12] "FAQ—Edition 2.0: E) Key concepts". IEC 61508—Functional Safety. International Electrotechnical Commission

[13] Sommerville I. Software Engineering. Boston, Massachusetts (USA): Pearson Education; 2015

[14] Horbach J, Rammer C, Rennings K. Determinants of eco-innovations by type of environmental impact—The role of regulatory push/pull, technology push and market pull. Ecological Economics. 2012;**78**:112-122

[15] Dietz T, Rosa E. Rethinking the environmental impacts of population. Affluence and Technology. Human Ecology Review. 1994;**1**(2):277-300

[16] HP Face-Tracking Webcams Don't Recognize Black People [Internet]. Gizmodo.com. 2019. Available from: https://gizmodo.com/hp-face-tracking-webcams-dont-recognize-

black-people-5431190 [Accessed: 12 April 2019]

[17] Bowles N. 'I think my blackness is interfering': Does facial recognition show racial bias? [Internet]. The Guardian. 2019. Available from: https://www.theguardian.com/technology/2016/apr/08/facial-recognition-technology-racial-bias-police [Accessed: 12 April 2019]

[18] Ralph P, Kelly P. The dimensions of software engineering success. In: Proceedings of the 36th International Conference on Software Engineering (ICSE); 2014; 2014

[19] Galhotra S, Brun Y, Meliou A. Fairness testing: testing software for discrimination. In: Proceedings of the 2017 11th Joint Meeting on Foundations of Software Engineering (ESEC/FSE) 2017; 2017

Chapter 4

Embedded Systems Based on Open Source Platforms

Zlatko Bundalo and Dusanka Bundalo

Abstract

Ways and possibilities for design, implementation and application of microcomputer-based embedded systems using open source hardware and software platforms are considered, proposed and described. It is proposed to use open source hardware and software microcomputer-based technologies for design and implementation of embedded systems in many practical needs and applications. Main advantages and possibilities of application and implementation of such embedded systems are considered and described. Two practically designed and implemented systems performing needed data acquisition and control are presented and described. Used technologies for realization of the systems and embedded applications of the solutions are described. Open source microcomputer boards, appropriate sensors, actuators and additional electronics are used for implementation of the systems hardware. Open source tools and programs and LINUX operating system are used for implementation of the systems software. Modular approach is applied in the systems design and realization. Easy system expandability, simplifying maintenance and adaptation of the system to user requirements and needs are enabled with such approach. Balance between functionality and cost of the systems was also achieved. Optimization according to user requirements and needs, low consumption of electrical energy and low cost of the system are main advantages of such systems compared with standard embedded systems. These systems are optimized and specialized systems for specific needs and requirements of users.

Keywords: microcomputer-based embedded systems, open source hardware and open source software, LINUX operating system, modular system design, low cost of system, low consumption of electrical energy, system specialization and optimization

1. Introduction

Rapid development of electronic, computer and information technologies has enabled design and implementation of different types of so called embedded systems used for many different needs and purposes, for many practical applications. Such systems are designed and implemented to be maximally adapted to concrete application and to efficiently perform specific needed function. Also, such systems should be as simple as possible, with as much small dimensions as possible, as cheaper as possible and to consume as little energy as possible. Those systems are always specialized and optimized systems for some specific purpose and are used in almost all areas of human activities [1–4].

Different hardware and software solutions, different hardware and software platforms, can be used for development, design and implementation of such embedded systems. Those are mostly microprocessor-based solutions and platforms. What practical solution will be used primarily depends on specific requirements that such system should satisfy, in accordance with concrete need. Specially designed solutions and specially designed platforms are most often used for such purposes [1–4]. But, it can be shown that in many practical needs it is much more convenient and better to use so called open source hardware and software solutions, platforms of open source type [1–10].

Possible ways of design, implementation and application of embedded systems based on open source microcomputer hardware and software platforms are proposed and described here. Using open source hardware and software microcomputer-based technologies for design and implementation of embedded systems in many practical needs and applications is proposed. Main possibilities and advantages of such embedded systems are considered and described. Two such practically designed and implemented systems that perform appropriate data acquisition and control are presented and described. The systems hardware was implemented using open source microcomputer boards, appropriate sensors, actuators and additional electronics. The systems software was implemented using open source tools and programs and LINUX operating system.

The remainder of this chapter is organized in a following way. Section 2 considers basic concepts of embedded systems, main characteristics, way for design and motivation for practical application and implementation of such systems. Section 3 describes main concepts of proposed embedded systems design and implementation based on open source platforms. This section proposes way of design and implementation of open source hardware and software-based microprocessor embedded systems as specialized systems optimized for concrete user need. It also describes advantages of such design and implementation. Section 4 describes design of one practically implemented open source-based microprocessor embedded system for user identification in access to objects using mobile phone, as specialized systems optimized for user need. Section 5 shows and in more details describes design and implementation of one such based SCADA and RFID embedded system for industrial process control application, implemented using LINUX platform. Finally, a short summary and conclusions are given in Section 6.

2. Embedded systems

Embedded systems are electronic systems with very strong integration of hardware and software, designed to perform some specific functions [1–4]. Those are computer-based electronic systems built into another system, for what they provide better functionality and performance. Embedded systems are special purpose systems where the system is fully encapsulated by the device it controls. Such system performs one or more predefined tasks, usually with very specific requirements. Since system performs specific task, designer can optimize it in speed of operation, also reducing system dimensions and weight, energy consumption and cost. Embedded systems are often products of mass production and mass consumption. Such systems use microcomputer-based implementation [1–4].

Some of the basic characteristics of embedded systems are: performing specific tasks, supported by wide range of processors, real time limitations and should be cheap. If operating system is used it usually is Real Time Operating System (RTOS). In many cases such systems are battery powered and low-power systems, often must operate in extreme ambient conditions, have system resources necessary only for concrete use. Such systems keep all object code in ROM and require use of special means and methods for design [1–4].

There is wide range of application areas of embedded systems. Such systems are generally very inexpensive and are used in almost every electronic product. Some of main areas of embedded systems application are: computer peripheral devices, automotive electronics, aircraft electronics, trains, telecommunications, medical systems, military applications, authentication systems, consumer electronics, fabrication equipment, smart buildings and robotics [1–4].

Hardware design of embedded system includes selection of processor and interface logic for connection with environment. It also requires achieving balance in design and decision what will be realized by hardware and what by software. Decision how to divide design into hardware and software part is a key element in design of embedded system.

Characteristic of embedded systems is that cross development is usually used in software development. Generally, the embedded system software is developed on one platform and is executed on another platform. Also, it is required to obtain as efficient as possible programs in machine language. Since compilers do not generate efficient machine codes, programming in assembler language is often used. But, embedded applications are becoming more complicated and there is need to use programming in high level programming languages. Generally, there is no ideal programming language for programming of embedded systems. The most often used programming languages in embedded systems are: C, C^{++}, Java and DFL [1–4].

Processors used in practice can be divided into two groups: general purpose processors and special purpose processors. In the initial period embedded systems were using general purpose processors. Development of VLSI IC technologies enabled that embedded systems are mainly realized using special purpose processors (so called embedded processors). A key characteristic of applying of such processors is that designer should well know nature of concrete application, to be able to meet requirements and to select appropriate processor for a system. There are different classes of embedded processors: microcontrollers, RISC processors, Digital Signal Processors (DSP), multimedia processors, Application Specific Instruction Set Processors (ASIP). There are also classes of such processors for achieving high performances in intensive computing applications. Some of typical such processors are: network processors, digital signal processors (DSP), SoC (System on Chip) processors [1–4]. Advantages of using general purpose processors are: low cost for research, design, develop and test, high flexibility, low time to market. Advantages of using special purpose processors are: fast, low power consumption and small size solutions [1–4].

Depending on specific application and required characteristics, and considering the platform used for development and implementation, there are practically possible three ways of implementation of embedded systems:

- Embedded systems based on specially designed platform,
- Embedded systems based on open source platform,
- Embedded systems based on PLC (Programmable Logic Controller) platform.

Embedded systems based on specially designed platform use specially designed hardware platform, designed only for concrete application. The platform can be based on general purpose microprocessors or on special purpose processors. FPGA-based processor platforms or SoC platforms are very often used. In design of the platform everything is optimized and minimized, in order to satisfy all needs of concrete application. As far as software platform is concerned in such systems, the assembler language for programming is mainly used. Most often, operating system is not used and all programs are application programs. Also, all software is minimized and optimized, with the goal to meet all application requirements. Standard hardware and software development tools, or even specially developed tools, are used for development of such systems. Advantages of such design and implementation of embedded systems are: minimal dimensions and weight, minimal energy consumption, minimal price, maximal speed of operation, full optimization and minimization of all hardware and software resources. Disadvantages of this approach are: very long development time and great development costs, long time to market. This method of design and implementation is used when it is needed to realize and produce very large number of such systems, in products of mass consumption and mass production. In such situations this method has the best economic effects and such are obtained the systems with the lowest market price.

Embedded systems based on open source platform use some of on the market available open source hardware platforms. Such platforms are based on general purpose microprocessor or microcontroller or on DSP processor. For design and implementation of concrete system it is needed to select open source hardware platform, among all available. That platform should be the simplest and the cheapest and should meet all requirements of application. Concrete platform is chosen such that entire system be optimized and minimized, but with the goal that all requirements for application are met. As far as the software platform is concerned, high level programming languages are mainly used for programming in such systems. Rarely is used assembly language. Often, operating system is used in such systems, usually RTOS. Entire software is minimized and optimized with the aim to satisfy concrete purpose. Standard hardware and software development tools are used for development of the systems. Advantages of such method of design and implementation are: smaller development time and lower development costs, less time to market, optimization of all resources, and relatively high speed of operation. Disadvantages of this method are: greater dimensions and weight, higher energy consumption, higher price. This method should be used when it is needed to realize and produce relatively smaller number of such systems. In such situations this way of design and implementation has the best economic justification and gives the best economic effects.

Embedded systems based on PLC platform use some of on the market available PLC devices as hardware platforms. Such platforms are based on general purpose microprocessor or microcontroller or on DSP processor used in some available PLC device. For design and implementation of concrete system it is needed to select PLC platform, among all available. That platform should be the simplest and the cheapest, and should meet all needs of application. Concrete PLC platform is selected such that entire system can be optimized and minimized, but with the goal that all requirements for system application are met. As far as the software platform is concerned, in such systems mainly are used tools and languages that are used for PLC programming. Whole software is minimized and optimized according to application and with the aim to satisfy purpose of the system. Standard resources and tools for development of PLC-based systems are used for development of such systems. Advantages of such way of design and implementation are: high speed of operation, high reliability, short development time and low development costs, short time to market. Disadvantages of this way are: large dimensions and weight, high energy consumption, high price. This method is used when is needed to realize and produce small number of such systems. In such situations this way of design and implementation has the best economic justification and gives the best economic effects.

3. Design of embedded systems based on open source platform

High performances, high speed of operation, minimal energy consumption and large number of such solutions are not required in many practical needs and applications of embedded systems. In such applications the most important is that all needed system functionalities are met, with as short as possible system development and implementation time, and as little as possible system development and implementation costs [1–10]. It is also required as much as possible optimization and minimization of hardware and software, to achieve needed speed of operation. Also, it is needed as little as possible dimensions, weight and energy consumption, as high as possible reliability, as easy as possible maintenance, and as low as possible price of the system. For such purposes the best is to use embedded systems based on open source platforms. Such way of design and implementation enables simplification and acceleration of development, design and implementation process, reduction of time needed for all the activities, and reduction of costs of all the processes. It also enables modular approach, what facilitates, accelerates and reduces costs of system modification and maintenance. It is possible to optimize all hardware and software resources in accordance with needs of user, and to balance all necessary functionalities and price of the entire system.

Use of open source hardware platforms (suitable open source boards) drastically reduces and makes easier and cheaper process of development, design and implementation of hardware and entire system. It is only necessary to choose and select the most suitable open source platform for concrete application in accordance with all requirements of that purpose. It is also needed to select some additional hardware modules that should be added to the selected open source platform. Such is enabled modular hardware approach if it is necessary for concrete application. Other activities in design of hardware include defining ways and means for connecting system with environment, with sensors and actuators, and with user. It is very simple to perform since available hardware open source platforms offer large number of ways and circuits for interconnection and communication. In such design, it is constantly intended to minimize and optimize entire system hardware, and to satisfy concrete application of the system. With this way of design and realization of system hardware, it is practically very possible and feasible, with very low total cost.

Use of open source software platforms (suitable open source software development tools) reduces and makes easier and cheaper process of development, design and implementation of software and entire system. The development tools include open source programming languages and software tools, operating systems and debugging tools. It is only necessary to select the most suitable open source software tools for concrete application. For programming it is mainly used and it is intended to use as much as possible the high level programming languages. Such development of programs is simplified, accelerated and cheaper. Problem could only be in applications where very high speeds of operation are required, where programming in high level programming language can not meet needed speed. Programming in assembler language is used in such situations. But, it complicates, slows down and increases cost of development of the program. Therefore, it is tried to use programming in high level programming languages wherever it is possible. Programming in assembler language is used only in those parts of program where is needed as high as possible speed. It is also necessary to implement software modules that could be used in design of complete system software. Such is enabled modular software approach if it is necessary in concrete application. Also, in such systems it does not have to be used operating system. But some open source operating system can be used. As operating system slows down operation of the system and requires more hardware resources (larger memory), it is not used in applications where high speed of operation and small hardware resources are needed. In applications where it is not critical, it is preferable to use some open source operating system. Use of operating system very much facilitates and accelerates software design, implementation, modification and maintenance in entire system. It simplifies, accelerates and makes cheaper all activities in development and implementation of software and entire system. Further activities in development of software include programming of system communication with environment and with user. It is simple in such systems since available open source software platforms offer large number of different programs (drivers) for such communication for used hardware open source platforms. It is also constantly intended to optimize entire software of system and to satisfy concrete application. With this way it is practically very possible and feasible, with very low total costs. Manufacturers of hardware open source platforms also offer appropriate software open source platforms for their products that are mainly used in software development of such systems. This significantly simplifies, accelerates and makes cheaper development, design, implementation and maintenance of software of such systems.

Optimal integration of hardware and software of entire system is also very important in design of embedded systems. That enables and ensures achieving optimal characteristics of the system with satisfying all requirements, and with balance of hardware and software of system. In design of system, it is needed to decide what will be realized by hardware and what by software. It is always needed to minimize use of hardware and to maximize use of software. As less as possible, functions should be realized by hardware and as much as possible by software. It enables minimization of dimensions, weight, energy consumption and cost of the total system. Therefore, in development and implementation of such embedded systems, from the beginning should be parallel and simultaneously realized design of system hardware and software and their integration and verification. That all is also simplified in such embedded systems since are mainly used open source hardware and software platforms of the same manufacturer.

4. Embedded system for user identification in access to objects using mobile phone

Mobile phones and smartphones are standard communication devices that are used by many users. Mobile phone is microprocessor-based system with different peripheral communication elements and memories. SIM (subscriber identity module) card or smart card in mobile phone can memorize many data about user of mobile phone and different identification data (user name, access passwords and other identification data). It enables that mobile phone could be also used as identification device. The mobile phone could consolidate or replace many identification elements, different types of cards and other elements for identification. Such identification elements are for example: identity card, health card, passport, driving license, student index, remote device for access to a car, remote device for access to some object or area, for entering public transport vehicles, for electronic payments, etc. All that identification elements could be replaced by one mobile phone and identification activities could be performed much simpler. The embedded system that enables mobile phone to be not only communication, but also identification device is proposed and described here. Using such approach and such systems it can be avoided usage of other identification elements in many applications with user identification, for monitoring, tracking and control of access of users to different facilities, spaces and services using a standard mobile phone.

Developed and implemented system is based on application of open source Arduino microcontroller platform and smartphone with open source Android operation system. Bluetooth Low Energy (BLE) wireless technology was used for communication [5, 6]. The implemented solution is used for control of access of user to objects, for control of user access to the car parking place or garage. Only users with needed identification data in their smartphone and with appropriate International Mobile Equipment Identity (IMEI) number of the smartphone can access to the car parking place or garage. The system is connected with computer of PC type that is used as a central station. It gives possibility to realize monitoring of user presence in the object and memorizing and processing of such obtained information and data. It can be connected with other central stations or with main monitoring centre in the case of more complex system realization.

In order to increase reliability of control it is needed authentication of the user [5–8]. There are practically three authentication methods based on: something that user knows (password, key word, etc.), something that user owns (key, token, etc.), something that user is (voice, fingerprint, etc.) [11]. Implemented system uses and combines the first two methods: something that user knows (user name and password) and something that user owns (smartphone with appropriate IMEI number) [12].

4.1 Smartphone in identification of users

One of the most important characteristics of the mobile phone is portability. It allows mobile phone to be permanently close to the user and to be such an ideal element for user identification. Hardware architecture of smartphone is different from architecture of standard general purpose processor. There are significant differences in architecture of CPU used in smartphone [13]. Smartphone uses operating system as an integrated element [13]. The most used operating systems in smartphones are: Google Android, Apple iOS, Nokia Symbian, RIM BlackBerry, Samsung Bada, Microsoft Windows Phone. The application is a part of smartphone software providing additional features of smartphone. By using applications user adds needed functionalities to smartphone.

Bluetooth Low Energy (BLE) and Near Field Communication (NFC) are wireless technologies often used for data transfer between smartphone and other systems [14, 15]. Designers should to select one of the technologies that meet their needs. The BLE wireless technology was used for practically implemented and described embedded system. Reduced power consumption and low cost are the most important advantages of BLE wireless technology. These advantages are reasons for selecting BLE wireless technology for communication between smartphone and Arduino board-based system in the implemented solution.

Users carry many identification cards and devices every day. These identification elements are of different shape and production. That complicates all user activities that require identification. Many of user identification elements could be integrated into mobile phone. It is easier for users to carry one mobile phone with identification data for all identification systems they use than to carry many of different elements for identification. The risk of forgery and malversation also decreases in case of losing or stealing of identification element with some kind of intelligence as mobile phone is.

4.2 Design and implementation of user identification system using mobile phone

Design and implementation of user identification system using smartphone for entrance to the garage or to the car parking space is proposed and described here [5, 6]. **Figure 1** shows structure of the designed and implemented embedded system.

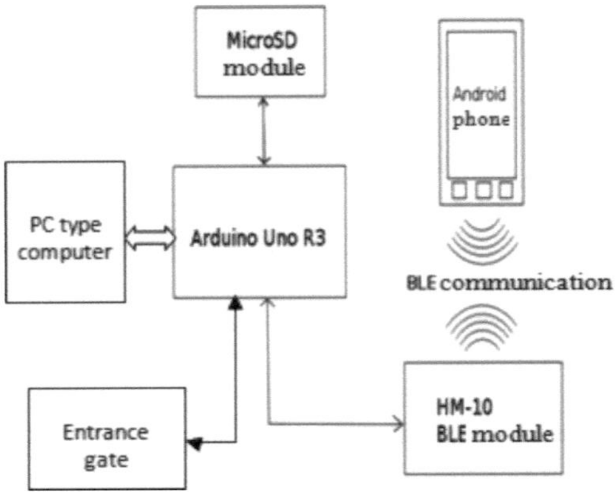

Figure 1.
Structure of implemented user identification system.

Smartphone with the Android operating system is used for entering user identification data at the entrance of the object (garage or car parking space). The microcomputer Arduino Uno R3 platform with ATmega328P microcontroller is used for checking and control of the identification process and for approval of user entry [15, 17]. The mechanism and elements for opening or closing entrance gate of object is connected to and controlled by the Arduino platform. Using Android application on the smartphone user enters username and password. That identification data is transferred via BLE wireless communication to the Arduino platform that is located in the object access point. Smartphone IMEI number is also transferred in the same time. User identification data including smartphone IMEI number is stored in the MicroSD card. Program on the Arduino platform checks out entered user data. User was successfully identified if sent user information is correct (matches with the user data stored on the MicroSD card). It is also obtained permission to access the garage or the car parking and entrance gate is opening. In the case that sent user information is not correct the access is not allowed. Arduino platform is connected with PC type computer that functions as system central station. Information about users, their permissions, accesses, presence and other data are memorized and kept on PC computer and can be analyzed on that computer. The PC type computer can be connected to other such computers (other central stations) and one main monitoring station if it is needed to realize more complex system for access control and monitoring more parking spaces or garages.

Valid user identification data is memorized on the MicroSD module. In that way was avoided entering identification data in Arduino program. It was also enabled to easily add, delete and update identification data using any device that can work with SD card (computer, mobile phone, etc.). Also, user identification would not be possible if the SD card is removed from SD module. It could be very simple mechanism to forbid further use of the system.

Communication and identification data exchange with Android-based smartphone is performed using HM-10 BLE wireless module. Two LEDs connected to Arduino board are used for indicating state of user identification and state of entry gate.

Open source Arduino IDE development environment was used for development of the embedded system [16, 17]. Arduino programming language very similar to C^{++} programming language was used for programming [16, 17]. Open source Android Studio was used for development of Android application [18].

Designed and implemented programs have two parts: program for the Arduino board and program for the Android smartphone. Program part for the Arduino board accepts commands from mobile phone and returns responses. All elements connected to the Arduino board are also controlled by that program. **Figure 2a** shows part of that program. Program part for the Android mobile phone processes answers obtained from the Arduino program and the Arduino board and interacts with the user. **Figure 2b** shows part of that program.

The user can use next command from Android mobile phone while the user is not identified and login:

- login—user identification process initialization (the LED 2 is turned on when user is identified).

The user can use the following commands after user identification:

- enabling access—command indicating that user was identified and enabling that access be enabled and open (open entry or access, this command turns on LED 1),
- disabling access—command indicating that user has left the space or service and that access be disabled and closed (closed entry or access, this command turns off LED 1),
- access status—command that returns information is entry enabled (open) or disabled (closed), that the LED 1 is on or off,
- logout—command by what the user interrupts and stops status of identified user and logout from the system (the LED 2 is turned off).

The Android application program on smartphone has three parts. The first part of the program scans whether active BLE devices exist. If exist active BLE devices near they will be shown in the list of devices to what the user can be connected. The second part of the application program is started when device from the list was selected and the connection process started by user pressing button "CONNECT". The user enters identification data (username and password) in this part of the

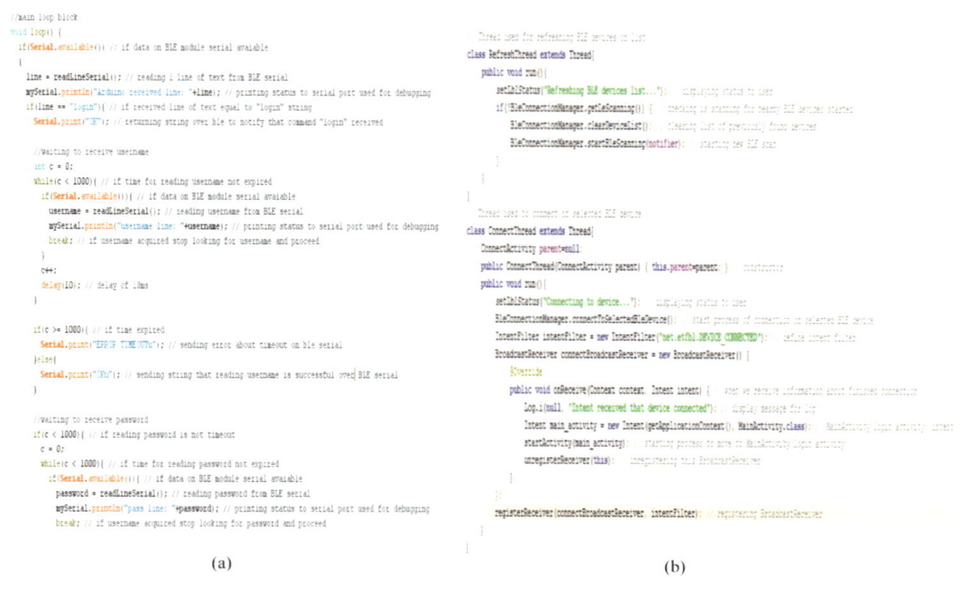

Figure 2.
Part of Arduino program (a) and part of android program (b).

program. Process of identification starts when user enters the data and presses the button "LOGIN". IMEI number of user smartphone is also transferred in the same time. User will receive message about wrong identification and will be forbidden system usage if the user enters wrong data or if IMEI number is wrong. The user is identified and goes to the third part of the application program if the transferred data is correct. The third part of the application program is the area where the user can stay or can leave the state of identified user and can return to the second part of the application program.

The application program on Android smartphone interacts with Arduino program part for identification. List of BLE devices to what users can be connected appears after starting Android application. It can be also restarted the scan process by pressing the button "REFRESH". It enables to find BLE devices that were later entered into the range of mobile phone or previously were not found for some reason (**Figure 3a**). By selecting one of the BLE devices from the list it is selected and marked that BLE device. Connection with the selected BLE device is performed by press on the button "CONNECT" (**Figure 3b**). It will be displayed the message "Please select device from list to connect" if the user was not selected any BLE device.

The application goes to login process and login form and starts identification of user after application connects to BLE device. **Figure 4a** shows that. User enters username and password in this form. By press on the button "LOGIN" it is performed sending of user identification data including user smartphone IMEI number to Arduino system (**Figure 4a**). User will receive an error message if sends wrong data (**Figure 4b**).

The user is login on the system and in Android application if the identification process was successful. LED diode (LED 2) connected to output of the Arduino board turns on (indication of user identification). User then goes on process and the form where can control entrance gate and LED diode (LED 1) connected to output of Arduino board (indication of access control). **Figure 4c** shows smartphone display form for the case when the user is identified and login.

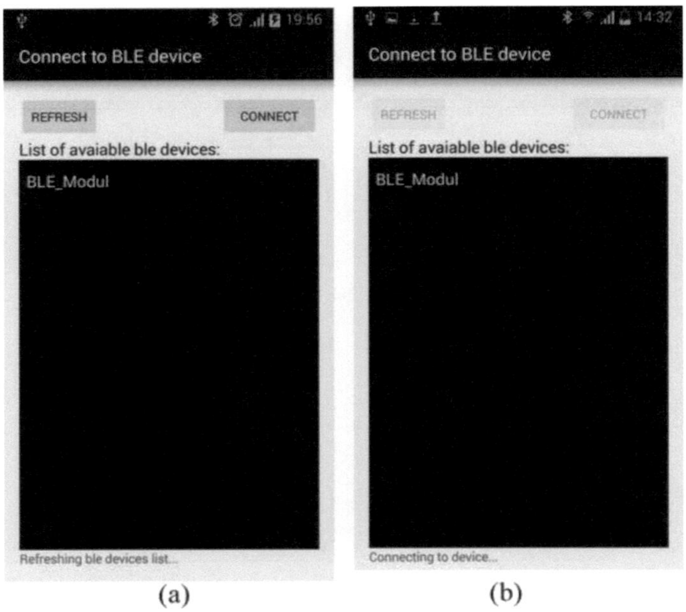

Figure 3.
Refresh of BLE devices list (a) and connection to selected BLE device (b).

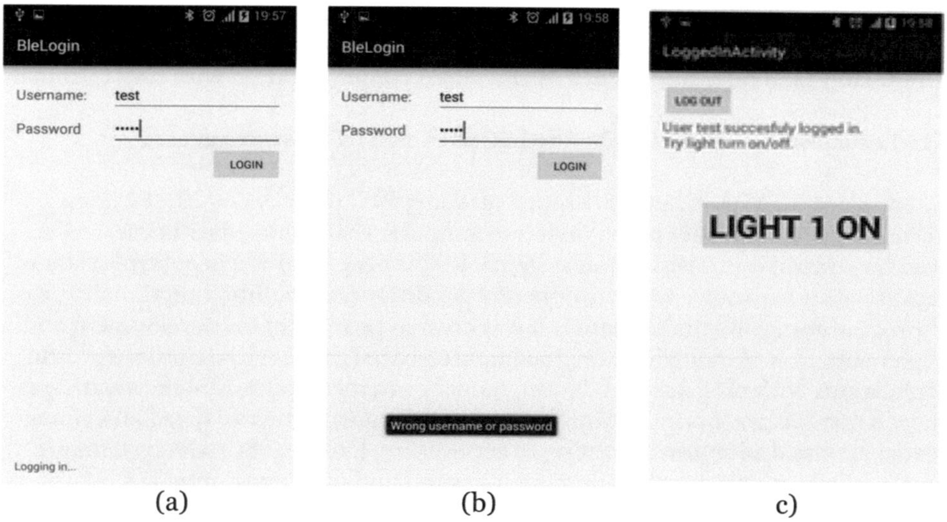

Figure 4.
User started with login process (a), user entered wrong user data (b) and login user form (c).

The implemented and described system shows that such way of use of embedded system, smartphone and wireless communication can very much simplify user daily and business activities and reduce costs. It can be simply enabled to access many places, objects and services and to perform complex operations and send and receive various data by using smartphone for identification. It is also used smartphone IMEI number to avoid misuse of the system by unknown and unwanted person.

5. Embedded SCADA and RFID system implemented on Linux platform

Microcomputer embedded systems are present in almost every part of human activities and environment. Application of such systems enables needed functionality, mobility and autonomy of many implemented systems with low cost. Some of many such systems applications are: acquisition, processing and storage of data from process; monitoring and control of processes; increasing security in the process. All this leads to so called Supervisory Control And Data Acquisition (SCADA) systems used for monitoring, control and data acquisition in industry systems [19, 20]. One of the most important requirements in industrial processes is security. So called Radio-Frequency Identification (RFID) is often used as security technology for identification of participants in the process [21, 22]. RFID is standard technology for identification objects and users in many cases because of contactless identification, relatively high speed, simplicity and low price.

Possibilities for design, implementation and application of embedded SCADA and RFID systems realized using open source technologies are considered, proposed and described here. One practically designed and implemented system for data acquisition and needed control in one industrial process is described [9, 10]. System hardware was implemented using open source microcomputer boards Pandaboard [23], Arduino/Genuino Uno and Mega [24], RFID sensors, temperature and air humidity sensors, WiFi router and additional electronics. Software was implemented using open source tools and programs on LINUX operating system [25]. Acquisition data of the system are: identification numbers of RFID sensor and RFID tag, temperature, air humidity, time of identification. Implemented system enables:

displaying information on LCD (Liquid Crystal Display), control using RFID, control and displaying of information using WUI (Web-based User Interface) and CLI (Command Line Interface), wireless and wired communication with user terminals.

5.1 Microcomputer based embedded SCADA and RFID systems

Standard SCADA systems are based on using PLC devices [19, 20]. PLC is a special type of controller based on microcomputer that works in real time and is used for control of processes and systems. PLC is very common in control systems because of its resistance to environmental conditions, reliability, functionality, ease of programming and maintenance. It was created possibility for development and implementation of embedded microcomputer-based SCADA systems for specific applications with progress and development of microcomputers. Main advantages of such systems are: optimization of system hardware, software, speed and storage needs; maximal adoption according to application, low cost. SCADA systems are consisted of: measurement equipment (sensors) and executive units (actuators), remote input/output modules, remote stations (terminal units), communication system and central station.

Radio-frequency identification is used in RFID systems [21, 22]. Process of identification is performed such that the RFID sensor detects the RFID tag brought in proximity to the sensor. It is performed identification of a person or of an object based on data in the tag. After successful identification it is enabled access of a user or an object to needed functions. RFID readers/writers use electromagnetic field for automatic detection and monitoring RFID tags. There are three types of RFID tags: passive RFID tags, active RFID tags and partly active RFID tags. Diversity of RFID tags enables selection of the most appropriate solution for a given application.

5.2 Design and implementation of SCADA and RFID system

The practically developed, designed and implemented system is specialized embedded SCADA and RFID system realized using open source hardware and software technologies [9, 10]. Implemented system enables:

- Control (turn on/turn off) of two high-power (energy) consumers using WUI, CLI and RFID sensors,
- Monitoring and changing description of the consumers,
- Acquisition and storage of information from RFID, temperature and air humidity sensors into data base,
- Two types of users (system administrator and ordinary users),
- Different authorizations for administrator and ordinary user,
- Authorization of user when accessing the system,
- Addition and change of users by administrator using WUI,
- Addition and change of RFID tags by administrator using WUI,
- Record of addition and change of users in the data base,

- Record of addition and change of RFID tags in the data base,
- Record of turning on and turning off of high-power consumers in the data base,
- Access to the system by SSH (secure shell) protocol using RSA cryptographic algorithm with key length of 2048 bits,
- Monitoring of the resources, network and processes of the system,
- Display of last 100 logs using WUI.

Process of design and implementation of the system was performed in next phases: hardware design, Bootloader design, Linux kernel and Device Tree design, ROOTFS and supporting applications, preparation of SD (Secure Digital) memory card, microcontroller programs design and programming. It was used one systematic approach for proper and optimized design of the system with the aim that the system performs all required functions and operations. Block scheme of designed and implemented system is shown in **Figure 5**.

The hardware in implemented system consists of open source elements: Pandaboard, Arduino/Genuino boards, LCD display, WiFi router, appropriate electronics and sensors.

The Pandaboard is open source development board based on dual-core 32-bit high-performance multimedia application microprocessor Open Multimedia Application Platform 4430 (OMAP 4430) and 1GB RAM and SoC (System on Chip) architecture [23]. Microprocessor OMAP 4430 is based on ARM (Advanced RISC Machine) Cortex-A9 architecture with very good performances and very small electrical energy consumption. The Pandaboard enables support for higher level operating systems and has performances that satisfy all needed requirements for central station in implemented SCADA and RFID system.

The Arduino/Genuino Uno and Arduino/Genuino Mega 2560 are open source development boards based on high performance, low power Atmel AVR 8-bit microcontrollers ATmega328P and ATmega2560, respectively, with supporting programs and libraries [24]. In the system design were used the Arduino/Genuino development boards to decrease time for performing needed operations and to increase possible distance between the system modules.

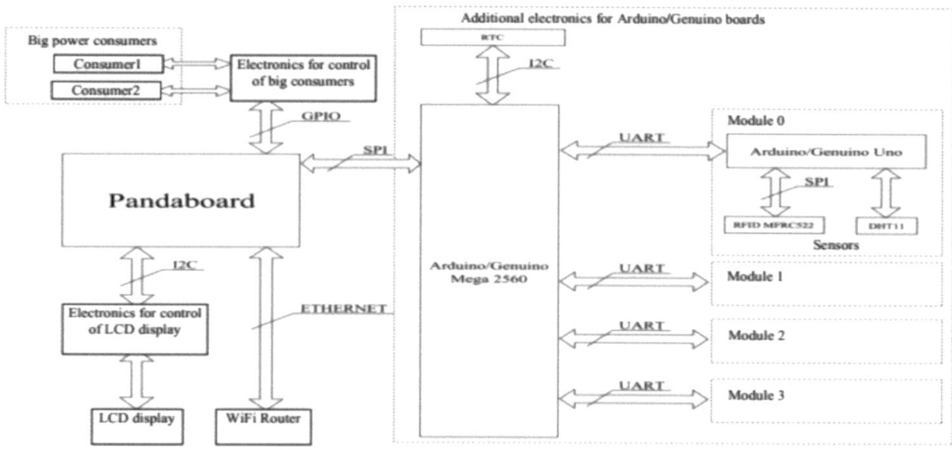

Figure 5.
Block scheme of implemented SCADA and RFID system.

Software of the system was implemented using open source software tools and programs. Appropriate open source tools and programs were used for development and implementation of software for Pandaboard and for Arduino/Genuino boards.

For implementation of hardware of the system were used: Pandaboard as central station, Arduino/Genuino boards as remote stations (system modules) and part for communication with the modules, additional electronics for the boards, electronics for control of high-power consumers, electronics for communication with LCD display, WiFi router TP-Link TL-WR740N. For implementation of additional electronics for Arduino/Genuino boards were used: Arduino/Genuino Mega 2560 board, I2C RTC (Real-Time Clock) based on DS3231 and AT24C32 circuits, modules.

Arduino/Genuino Mega 2560 board realizes communication of remote modules with the Pandaboard as central station. Such is enabled communication of the Pandaboard with four modules. Total number of remote stations in the system can be increased using multiplexors. The RTC circuit that provides information about the exact time even in a situation when the entire system is left without the main power supply source is connected on the Arduino/Genuino Mega 2560 board. Such operation is enabled by a battery integrated into the RTC circuit.

For implementation of the module (remote station) were used: Arduino/Genuino Uno board; temperature and air humidity sensor DHT11; RFID sensor MFRC522. The remote station enables system management and control using RFID tags from remote locations. It also sends needed information from the sensors to central station Pandaboard that has to perform appropriate actions on the complete system. This solution was selected to enable as much as possible system modularization. This way of design also enables easier service, maintenance and diagnostics of the errors in the system. It gives possibility for quick fault detection and replacement or repair of faulty module in event of a malfunction in the system. The system modularity results in higher reliability and greater configurability of the systems. So, it is faster and easier to customize and optimize the system in accordance with needs of user. Scheme of the module and scheme of additional electronics for Arduino/Genuino boards is shown in **Figure 6**.

Electronics for control of high-power consumers was implemented using: relays, Darlington transistor array circuit ULN2001A, resistors, LED diodes. There are

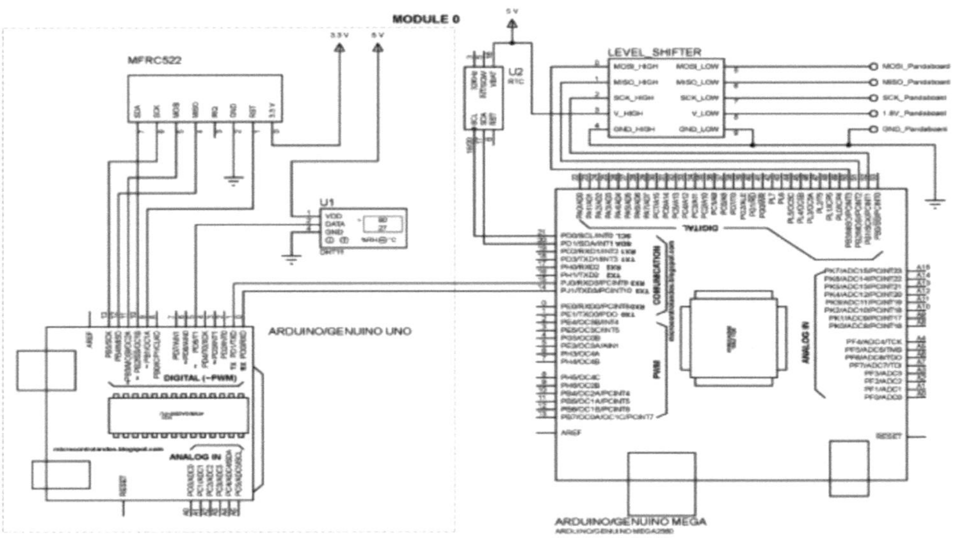

Figure 6.
Scheme of module and additional electronics for Arduino/Genuino boards.

two high-power consumers in the system. The LED diodes indicate state of the high-power consumers, what the high-power consumer is turned on or turned off.

Elements for turning on and turning off high-power consumers do not exist in the Pandaboard. The board also can not provide sufficient current and voltage for turning on and turning off the relays needed for control of high-power consumers. So, it was necessary to design additional electronics for control of the consumers. Integrated circuit ULN2001A with the Darlington coupling of transistors and with protection diodes was used in design of the scheme. The output current that the ULN2001A integrated circuit provides is 500 mA per channel. That is enough to ensure proper operation of the relays and indicator LED diodes. For protection of the system, it is needed that the power supply of electronics for control of high-power consumers be separated from the power supply of the Pandaboard. Such it is prevented that in a case of destruction of electronics for control of high-power consumers also be destroyed the Pandaboard, and vice versa. Designed electronics for control of high-power consumers enables the users to control the consumers with characteristics of AC (0–250)V and (0–10)A or DC (0–30)V and (0–10)A.

For implementation of electronics for control of LCD display were used: voltage level shifter, I2C communication module, LCD display and additional circuits. The voltage level shifter enables communication between Pandaboard and I2C communication module and LCD display that use different power supply voltage levels. The power supply voltage of the Pandaboard is 1.8 V and the power supply voltage for the voltage level shifter, I2C communication module and LCD display is 5 V. The maximal output current of the I2C pin of the Pandaboard is 3 mA. It is sufficient for correct operation of the voltage level shifter. Integrated circuit PCF8574 was used for I2C communication and LCD display control.

Development and design of needed electronics was performed using software Proteus 8 Professional on Microsoft Windows 10 operating system. After the design, it was practically implemented hardware part of the system that was tested. It was performed additional experimentation and testing of the system hardware part in order to decrease the system energy consumption.

The system module is remote station and operates as remote station. Communication of the module with the Arduino/Genuino Mega 2560 board is performed using UART interface. GPIO pins of the Pandaboard are used for communication of Pandaboard and electronics for control of high-power consumers. Additional electronics for Arduino/Genuino boards communicates with Pandaboard using SPI interface.

The Pandaboard is central unit and central station of the implemented system. It was used WiFi router to enable connection to the Pandaboard using mobile phone, tablet, laptop and other computers. Communication of the router and the Pandaboard is performed using Ethernet protocol. So, there are enabled WiFi and Ethernet connection to the implemented system. There is possibility to disable one of the two connections with the aim to increase communication security. WPA2 security protocol is used for protection of WiFi connection.

Development and cross compiling of the system software on the development computer was started after successful testing of the system hardware. The software cross compiling was performed using the software tool arm-linux-gnueabi-*.

Patch and source code of U-Boot bootloader v2015.10 downloaded from the Internet were used for design and implementation of Bootloader. The archived source code of the U-Boot and patch were downloaded from the Internet. In the design and implementation were performed: copying of patch into source code of U-Boot, setting parameters needed for cross compiling of U-Boot and patching of source code, deleting of not needed files and folders generated by previous compiling, cross compiling, copying of cross compiled U-Boot.

Version 4.1.14 of LINUX kernel downloaded from the Internet was used for design and implementation of the system. It was performed: programming of Device Tree file that enables correct mapping of devices during loading of LINUX kernel, setting parameters needed for cross compiling of LINUX kernel and Device Tree file, configuring of LINUX kernel, cross compiling of LINUX kernel and Device Tree file, setting of appropriate privileges to created files and folders.

During design and implementation of ROOTFS of the system first was cross compiled Busybox. It was used Busybox v1.24.1. and source code taken from the Internet. The archived source code was downloaded from the Internet and uploaded to the development computer. It was performed: setting variables needed for cross compiling of Busybox, configuration and cross compiling of Busybox, copying of needed libraries. In design and implementation of basic ROOTFS and configuring of the network were performed: creation of needed folders, creation of console devices, creation of ROOT users, setting of the network, copying of the files. During creation of ROOTFS it was used Dropbear version 2016.73 for SSH communication. It was realized: setting of variables needed for cross compiling of Dropbear, configuration and cross compiling of Dropbear, copying of needed files and folders, enabling automated starting of Dropbear when started operating system of Pandaboard.

SQLite program was used for management of data base. In the design and implementation were realized: setting variables needed for cross compiling of SQLite, configuration and cross compiling of SQLite, copying of needed files. **Figure 7** shows scheme of used data base. Information about users and RFID tags is stored and memorized in the data base. There are records about two types of users, the ordinary user and the administrator. Administrator can add new users and RFID tags, change activity state and update existing users and RFID tags. Every valid RFID identification is recorded in data base and information from sensors in that moment. For every change in data base is recorded time of change.

The Lighttpd, optimized open source HTTP server, was used for design and implementation of HTTP server. In the server design and implementation was realized: setting variables needed for cross compiling of Lighttpd, configuration and cross compiling of Lighttpd, creation of folder and copying of needed files, setting privileges to created folders and files. It was used Libxml2 library for parsing XML (Extensible Markup Language) and needed for cross compiling of PHP (Personal Home Page) used in design and implementation of web page. It was realized: setting variables needed for cross compiling of Libxml2, configuring and cross compiling of Libxml2. The PHP server-side script language was used for design and implementation of Internet web page. It was performed: setting variables needed for cross compiling of PHP, configuration of PHP, cross compiling of PHP, copying of files, creation of folder.

Application for control of high-power consumers was optimized and implemented as ash script to occupy minimum of space in implementation. Possibilities of the application are: turn on or turn off of consumer, monitoring state of consumer. Commands for control of consumers are: consumer turn on, consumer turn off, reading state of consumer.

Application for control of LCD display was also based on ash script. Command for control of LCD display gives possibility to show on the LED display needed string of text characters.

Application for communication with remote stations was written in C language. It enables acquisition of data from remote stations and recording important events into SQLite data base, and also time synchronization obtained from RTC.

It was designed and implemented graphical user interface in the form of Internet web page for monitoring and control of the system from remote locations.

DOI: http://dx.doi.org/10.5772/intechopen.85806

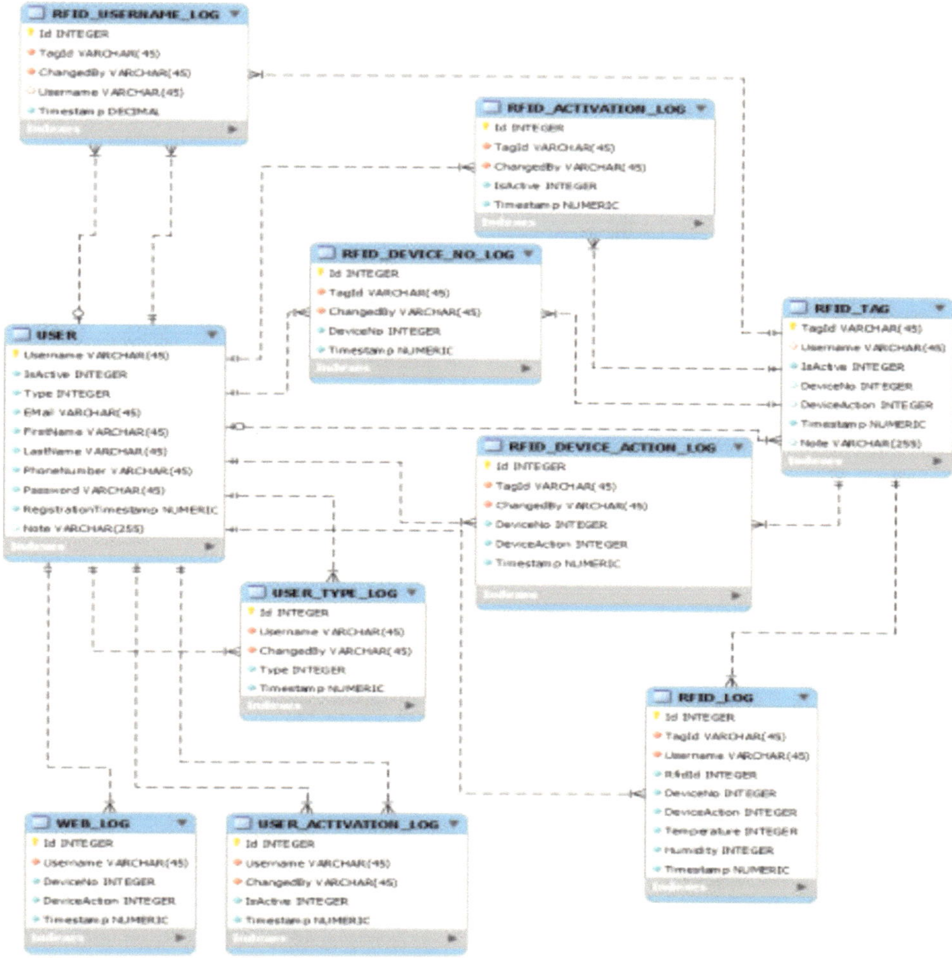

Figure 7.
Scheme of used data base.

For design and implementation were used next technologies: HTML (HyperText Markup Language), CSS (Cascading Style Sheets), PHP, JavaScript, AJAX (Asynchronous JavaScript And XML).

It was created and designed monitoring and control panel named PandaPower. It is needed to enter Username and Password on login page shown in **Figure 8a** during login to the system. It is not stored user Password in the data base but the hash of the user Password. In this way the administrator can not see the user Password and is prevented misuse of the data from the administrator side.

After authorization the PandaPower panel is available for the user. There are panels for ordinary user and for administrator. **Figure 8a** shows panel for ordinary user and **Figure 9a** shows panel for administrator. Ordinary user can use the panel for: Control of consumers and displaying load of SCADA system, Monitoring of consumers, network, processes and basic information of SCADA system, Displaying of the last 100 web logs. Administrator can use the panel for: Control of consumers and displaying load of SCADA system, Monitoring of consumers, network, processes and basic information of SCADA system, Displaying of the last 100 web and RFID logs, Control of RFID tags, Control of users. Every control of consumers, users and RFID tags is recorded in the data base. It is enabled to users to logout from the system after using of the PandaPower panel.

a) b)

Figure 8.
PandaPower login page (a) and panel for ordinary user (b).

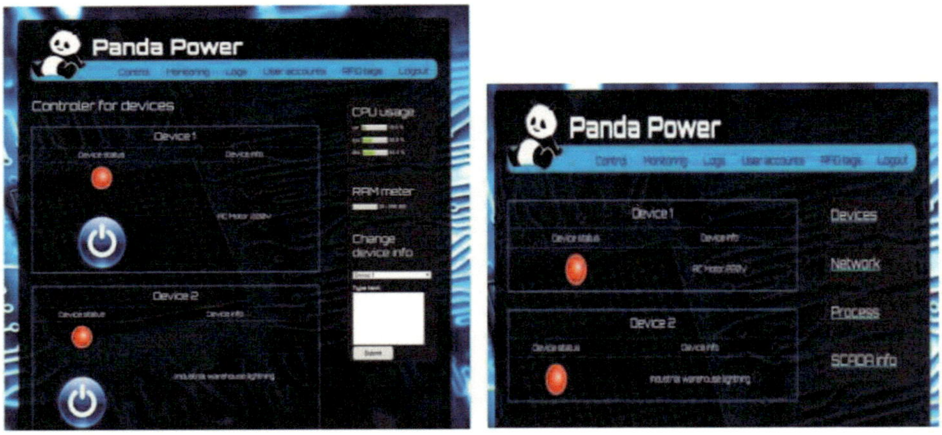

a) b)

Figure 9.
Panel for administrator (a) and administrator monitoring panel (b).

Administrator has possibilities to monitor and control users, consumers, network, process, RFID devices, basic information of the SCADA system. PandaPower administrator panel for monitoring consumers, network, process and basic information of the SCADA system is shown in **Figure 9b**. **Figure 10a** shows PandaPower administrator log panel that enables list of last 100 web and RFID system logs. **Figure 10b** shows administrator RFID panel for control of RFID tags.

Due to the support of Pandaboard, it was used SD memory card for storage of the data. SD memory card has appropriate performances and low cost. Formatting of SD card was performed on development computer using program fdisk on GNU/Linux Mint 17.2 operating system. Two partitions with names BOOT and ROOTFS were created on the SD card. Then were copied LINUX kernel, Device Tree, U-Boot and ROOTFS with all needed applications to the SD card.

It was also needed to program Atmel microcontrollers in the Arduino/Genuino boards. The Arduino IDE environment was used for the programming [24].

a) b)

Figure 10.
Administrator log panel (a) and administrator RFID panel (b).

Proposed and described embedded system is relatively complex but inexpensive. With such way of design the time needed for the system design and implementation, and system cost were significantly decreased.

6. Conclusions

Development of digital electronic technologies enables design and application of embedded systems for different needs and applications. Such systems are maximally adapted to concrete application and are specialized and optimized for a specific purpose. Different hardware and software platforms can be used for design and implementation of such systems. Specially designed platforms are the most often used for such purposes. But, in many practical needs it is much more convenient to use open source hardware and software platforms.

It is proposed here to use open source hardware and software microcomputer-based technologies for design and implementation of embedded systems for many practical applications. As illustration of such approach and its benefits, two such practically designed and implemented embedded systems were presented and described. Open source microcomputer boards, appropriate sensors, actuators and additional electronics were used for the systems hardware implementation. Open source software tools and programs and LINUX operating system were used for the systems software implementation. Modular approach that enabled easy expandability, simplifying maintenance and adaptation of the system to user needs was also used in the system design. It was also achieved balance between functionality and cost of the systems.

It was performed optimization in selection of all used components and tools, and optimization of all hardware and software designs, according to system needs, in the design and implementation of this embedded systems. Such it is achieved balance of functionalities and cost of the system. Main advantages of this embedded systems compared with other ways of design and realization are: lower cost, optimization according to application, smaller dimensions and weight, and lower electrical energy consumption. Some disadvantages are: lower reliability, lower certainty, constrained application on need for what the system was designed.

The Arduino hardware platform was chosen for the systems realization because that platform is easy available, very good documented, with many development tools, very cheap and can satisfy all needs of the systems concrete application with the small system cost. It is especially case with the system for user identification using smartphone that is relatively simple system. The Linux operating system was used in implementation of SCADA and RFID system since that system is more complex and not requiring high speed of operation. So, LINUX operating system was chosen because it enables faster and simpler design of system software, shorter system development time and smaller cost of the system.

Embedded systems based on open source platforms are not standard general purpose solutions or totally specially designed solutions. Such systems are less expensive, specialized and optimized solutions for some concrete user need. Such design and implementation should be used mainly for systems of smaller or medium complexity and where is needed smaller number of the systems. That approach can also be used for design and implementation of more complex and larger embedded systems. But, then it will be needed much more activities and time for system design and implementation and systems will be more expensive.

Author details

Zlatko Bundalo[1]* and Dusanka Bundalo[2]

1 Faculty of Electrical Engineering, University of Banja Luka, Banja Luka, Bosnia and Herzegovina

2 Faculty of Philosophy, University of Banja Luka, Banja Luka, Bosnia and Herzegovina

*Address all correspondence to: zlatbun2007@gmail.com

References

[1] Wolf W. Computers as Components: Principles of Embedded Computing Systems Design. San Francisco, USA: Morgan Kaufman; 2008

[2] Gajski D, Abdi S, Gerstlauer A, Schirner G. Embedded System Design: Modeling, Synthesis and Verification. Heidelberg, Germany: Springer; 2009

[3] Valvano J. Embedded microcomputer systems: Real-time interfacing. Boston; USA: Cengage Learning. 2011

[4] White E. Making Embedded Systems: Design Patterns for Great Software. Boston, USA: O'Reilly Media; 2011

[5] Bundalo Z, Ristić N, Bundalo D, Pašalić D, Cvijić B. Possibilities of using mobile phone for user identification. In: Proceedings of International Scientific Conference ERK2016; Portorose, Slovenia. 2016. pp. A:39-A:42

[6] Bundalo Z, Ristić N, Bundalo D, Pašalić D, Cvijić B. Possibilities of using mobile phone for user identification. ANNALS of Faculty Engineering Hunedoara—International Journal of Engineering. 2018;**16**(2):185-188

[7] Pašalić D, Cvijić B, Bundalo D, Bundalo Z, Kuzmić G. Embedded systems for user identification in access to objects and services using mobile phone. In: Proceedings of International Conference MECO2017; Bar, Montenegro. 2017. pp. 109-112

[8] Zulić M, Bundalo Z, Bundalo D. Microcontroller based embedded system for garage access control by mobile phone. In: Proceedings of International Scientific Conference ERK2017; Portorose, Slovenia. 2017. pp. 19-22

[9] Papić M, Bundalo Z, Bundalo D, Stojanović R, Kovačević Ž, Pašalić D, et al. Microcomputer based embedded SCADA and RFID systems implemented on LINUX platform. In: Proceedings of International Conference MECO2017; Bar, Montenegro. 2017. pp. 104-108

[10] Papić M, Bundalo Z, Bundalo D, Stojanović R, Kovačević Ž, Pašalić D, et al. Microcomputer based embedded SCADA and RFID systems implemented on LINUX platform. In: Microprocessors and Microsystems. Vol. 63. Amsterdam, Netherlands: Elsevier; 2018. pp. 116-127

[11] NIST Computer Security Handbook. 2008. Available from: http://www.sos.cs.ru.nl/applications/courses/security2008/nistiadraft.pdf

[12] Official Web site of GSMA. 2018. Available from: http://www.gsma.com/managedservices/mobile-equipment-identity/about-imei/

[13] Singh M, Jain M. Evolution of processor architecture in mobile phones. International Journal of Computer Applications. 2014;**90**(4):34-39

[14] Townsend K, Cufí C, Davidson R. Getting Started with Bluetooth Low Energy. Boston, USA: O'Reilly Media; 2014

[15] Igoe T, Coleman D, Jepson B. Beginning NFC: Near Field Communication with Arduino, Android, and PhoneGap. Boston, USA: O'Reilly Media; 2014

[16] Schmidt M. Arduino A Quick Start Guide. Dallas, USA: Pragmatic; 2011

[17] Blum J. Exploring Arduino: Tools and Techniques for Engineering Wizardry. Indianapolis, USA: John Wiley & Sons; 2013

[18] Annuzzi J, Darcey L, Conder S. Introduction to Android Application Development. Boston, USA: Addison-Wesley; 2013

[19] Bailey D, Wright E. Practical SCADA for Industry. Amsterdam, Netherlands: Newnes and Elsevier; 2003

[20] Knapp E. Industrial Network Security: Securing Critical Infrastructure Networks for Smart Grid, SCADA, and Other Industrial Control Systems. Amsterdam, Netherlands: Elsevier; 2011

[21] Finkenzeller K. RFID Handbook: Fundamentals and Applications in Contactless Smart Cards, Radio Frequency Identification and Near-Field Communication. Indianapolis, USA: Wiley & Sons; 2010

[22] Prabhu N. Design and Construction of an RFID-Enabled Infrastructure—The Next Avatar of the Internet. Boca Raton, USA: CRC Press; 2014

[23] OMAP4460 Pandaboard ES System Reference Manual. Available from: https://www.cs.utexas.edu/~simon/378/resources/PandaBoardES.pdf

[24] Web site of Arduino. Available from: http://www.arduino.cc/

[25] Simmonds C. Mastering Embedded Linux Programming. Birmingham, United Kingdom: Packt Publishing; 2015

Section 3

Algorithms

Chapter 5

The K-Means Algorithm Evolution

Joaquín Pérez-Ortega, Nelva Nely Almanza-Ortega, Andrea Vega-Villalobos, Rodolfo Pazos-Rangel, Crispín Zavala-Díaz and Alicia Martínez-Rebollar

Abstract

Clustering is one of the main methods for getting insight on the underlying nature and structure of data. The purpose of clustering is organizing a set of data into clusters, such that the elements in each cluster are similar and different from those in other clusters. One of the most used clustering algorithms presently is K-means, because of its easiness for interpreting its results and implementation. The solution to the K-means clustering problem is NP-hard, which justifies the use of heuristic methods for its solution. To date, a large number of improvements to the algorithm have been proposed, of which the most relevant were selected using systematic review methodology. As a result, 1125 documents on improvements were retrieved, and 79 were left after applying inclusion and exclusion criteria. The improvements selected were classified and summarized according to the algorithm steps: initialization, classification, centroid calculation, and convergence. It is remarkable that some of the most successful algorithm variants were found. Some articles on trends in recent years were included, concerning K-means improvements and its use in other areas. Finally, it is considered that the main improvements may inspire the development of new heuristics for K-means or other clustering algorithms.

Keywords: clustering, K-means, systematic review, historical developments, perspectives on clustering

1. Introduction

The accelerated progress of technology in recent time is fostering an important increase in the amount of generated and stored data [1–4] in fields such as engineering, finance, education, medicine, and commerce, among others. Therefore, there is justified interest in obtaining useful knowledge that can be extracted from those huge amounts of data, in order to help making better decisions and understanding the nature of data. Clustering is one of the fundamental techniques for getting insight on the underlying nature and structure of data. The purpose of clustering is organizing a set of data into clusters whose elements are similar to each other and different from those in other clusters.

One of the clustering algorithms more widely used to date is K-means [5], because of its easiness for interpreting its results and implementation. Another factor that has contributed to its use is the existence of versions implemented in the Weka and SPSS platforms and open-source programming languages such as Python and R, among others.

It is convenient to point out that K-means is a family of algorithms that were developed in the 1950s as a result of independent investigations. These algorithms have in common four processing steps, with some differences in each step. It was in an article by MacQueen [6] where the name K-means was coined.

The solution to the K-means clustering problem is hard, and it has been proven that it is NP-hard, which justifies the use of heuristic methods for its solution. According to the *no free lunch* theorem, there is no algorithm that is superior to other algorithms for all types of instances of an NP-complete problem. This has limited the design of a general algorithm for the clustering problem. For more than 60 years, a large number of variants of the algorithm have been proposed. There exist some surveys on K-means and its improvements. Classical surveys are [7] that synthesize K-means variants and their results, and in [8] a historical review is presented of continuous and discrete variants of the algorithm. In [9] several clustering methods and key aspects on clustering algorithm design are summarized, and a remarkable list of challenges and research directions on K-means was proposed. In [10] a review of theoretical aspects on K-means and scalability for Big Data is presented. Unlike these surveys, this documentary research focuses on classifying the K-means improvements according to the algorithm steps. This classification is particularly useful for designing versions customized of K-means for solving certain types of problem instances. This is also a contribution to the knowledge on the most important improvements for each step and, in general, to the behavior of the algorithm.

For selecting the most relevant articles, systematic review methodology was used. The filters used and the analysis of results allowed finding some of the most successful and referenced algorithm variants for each step of the algorithm.

The chapter is organized as follows: Section 2 summarizes the pioneering works that originated the family of K-means-type algorithms; additionally, it describes the standard algorithm and the formulation of the clustering problem. Section 3 describes the application of systematic review methodology for retrieving the most important articles on K-means; it also includes the step or steps to which the improvements apply and tables that summarize the number of articles; lastly, it includes a subsection on the new trends on the use of K-means. Finally, Section 4 includes the conclusions, highlighting the most successful and referenced algorithm variants.

2. Origins of the family of K-means-type algorithms

During the decades of the 1950s and 1960s, several K-means-type algorithms were proposed. These proposals were developed independently by researchers from different disciplines [8]. These algorithms had in common a process that originated what is currently known as the K-means algorithm.

According to the specialized literature [6, 11–20], four algorithms gave origin to this family. The following subsections describe the articles related to these algorithms and their authors.

2.1 Steinhaus (1956)

Mathematician Hugo Steinhaus, from the Mathematics Institute of the Polish Academy of Sciences, published an article titled "Sur la Division des Corps Matériels en Parties" [11], in which he presented the problem of partitioning a heterogeneous solid by the adequate selection of partitions. He also mentioned applications in the fields of anthropology and industry. Steinhaus was the first researcher that proposed explicitly an algorithm for multidimensional instances.

2.2 Lloyd (1957)

Stuart Lloyd, from Bell Laboratories, in the article titled "Least Squares Quantization in PCM" [12] approached the problem of transmitting a random signal X in a multidimensional space. Lloyd worked in the communications and electronics fields, and its algorithm is presented as a technique for pulse-code modulation.

2.3 MacQueen (1967)

James MacQueen, from Department of Statistics of the University of California, in his article titled "Some Methods for Classification and Analysis of Multivariate Observations" [6], proposed an algorithm for partitioning an instance into a set of clusters whose variance was small for each cluster. The term K-means was coined by him; it was known by different names: dynamic clustering method [13–15], iterative minimum-distance clustering [16], nearest centroid sorting [17], and h-means [18], among others.

2.4 Jancey (1966)

Jancey, from the Department of Botany, School of Biological Sciences, University of Sydney, in one of his articles titled "Multidimensional Group Analysis" [19], presented a clustering method for characterizing species *Phyllota phylicoides*. Jancey conducted his research in the field of taxonomy. There exists a variant of this method with similar characteristics, which was introduced by Forgy in the article "Cluster Analysis of Multivariate Data: Efficiency Versus Interpretability of Classification" [20]. The fundamental difference with respect to Jancey's work lies in the way in which the initial centroids are selected.

Because the results from Jancey's research will be used as reference for this chapter, his algorithm will be described in detail. The author stated that the similarity measures are based on the results published by the following authors: (a) Pearson in his article titled "On the Coefficient of Racial Likeness [21] published in 1926," (b) Rao in the article named "The Use of Multiple Measurements in Problems of Biological Classification" [22] published in 1948, and (c) Sokal in his article titled "Distance as a Measure of Taxonomic Similarity" [23] in 1961.

Pearson [21] in his article "On the Coefficient of Racial Likeness," when studying craniology and physical anthropology, confronted the difficulty of comparing two types of races, in order to determine the membership of a limited number of individuals to one race or another or both. As a result, Pearson proposed a coefficient of racial likeness (CRL). For calculating this coefficient, it is necessary to obtain first the means and variances of each characteristic in each sample, since it is assumed that there is variability for each of the characteristics considered. This coefficient is used to measure the dispersion around the mean and the degree of association between two variables.

The article published by Radhakrishna Rao [22] in the *Journal of the Royal Statistical Society*, titled "The Utilization of Multiple Measurements in Problems of Biological Classification," aimed at presenting a statistical approach for two types of problems that arise in biological research. The first deals with the determination of an individual as member of one of the many groups to which he/she possibly might belong. The second problem deals with the classification of groups into a system based on the configuration of their different characteristics.

Sokal [23] published his article titled "Distance as Measure of Taxonomic Similarity," which is based on the methods for quantifying the taxonomy classification process, and he points out the importance of having fast processing and data calculation methods. The purpose of his work is to evaluate the similarities among taxa that have observed characteristic values, instead of phylogenetic speculations and interpretations.

The similarity among objects is evaluated based on many attributes, and all the attributes are considered as equal taxonomic values; therefore, an attribute is not weighted more or less than any other.

For performing the weighting of attributes, three types of coefficients are used: association, correlation, and distance, where the last one is of interest for this study. This distance coefficient is employed for determining the similarity between two objects by using a distance function in an n-dimensional space, in which the coordinates represent the attributes.

A measure of similarity between the objects 1 and 2 based on two attributes would be the distance in a two-dimensional space (i.e., a Cartesian plane) between the two objects. This distance $\delta_{1,2}$ can be easily calculated through the well-known formula from analytic geometry, Eq. (1):

$$\delta_{1,2} = \sqrt{(X_1 - X_2)^2 + (Y_1 - Y_2)^2} \quad (1)$$

where X_1 and Y_1 are the object 1 coordinates and X_2 and Y_2 are the object 2 coordinates.

Similarly, when three attributes are needed for two different objects, it is now necessary to carry out the distance calculation in a three-dimensional space so that the exact position of the two objects can be represented regarding the three attributes. For calculating the distance between these two objects, an extension to the three-dimensional space of the formula for $\delta_{1,2}$ can be applied. When more than three dimensions are needed for the objects, it is not possible to represent their positions using conventional geometry; therefore, it is necessary to resort to algebraic calculation of data. However, the formula for distance calculation from analytic geometry is equally valid for an n-dimensional space.

The general formula for calculating the distance for two objects with n attributes is shown in Eq. (2):

$$\delta_{1,2}^2 = \sum_{i=1}^{n} (X_{i1} - X_{i2})^2 \quad (2)$$

where X_{ij} is the value of attribute i for object j (j = 1, 2).

Once the object classification process is completed, then the matrix of similarity coefficients obtained (based on object distances) can be used in the usual methods for clustering analysis.

Finally, it is important to emphasize the feasibility of calculating distance as the summation of the squared differences of the attribute values of objects of different kinds.

The clustering method proposed by Jancey consists of the following four steps:

1. Initialization. First, k points are randomly generated in the space, which are used as initial centroids.

2. Classification. The distances from all the objects to all the centroids are calculated, and each object is assigned to its closest centroid.

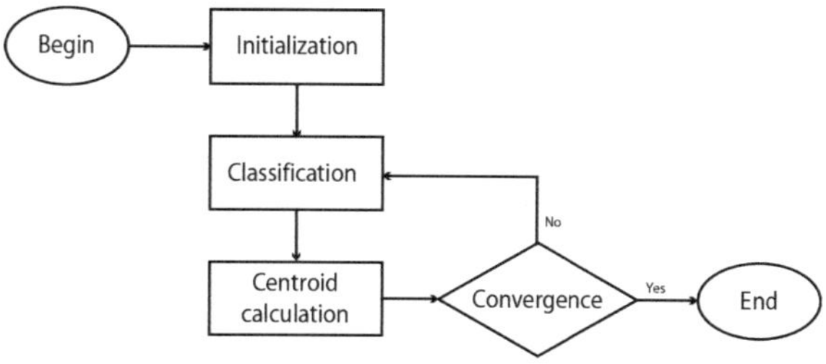

Figure 1.
Standard K-means algorithm.

3. Centroid calculation. New centroids are calculated using the mean value of the objects that belong to each cluster.

4. Convergence. The algorithm stops when equilibrium is reached, i.e., when there are no object migrations from one cluster to another. While no equilibrium is reached, the process is repeated from step 2.

2.5 *K*-means algorithm

K-means is an iterative method that consists of partitioning a set of n objects into $k \geq 2$ clusters, such that the objects in a cluster are similar to each other and are different from those in other clusters. In the following paragraphs, the clustering problem related to *K*-means is formalized.

Let $N = \{x_1, \ldots, x_n\}$ be the set of n objects to be clustered by a similarity criterion, where $x_i \in \Re^d$ for $i = 1, \ldots, n$ and $d \geq 1$ is the number of dimensions. Additionally, let $k \geq 2$ be an integer and $K = \{1, \ldots, k\}$. For a k-partition, $P = \{G(1), \ldots, G(k)\}$ of N, let μ_j denote the centroid of cluster $G(j)$, for $j \in K$, and let $M = \{\mu_1, \ldots, \mu_k\}$ and $W = \{w_{11}, \ldots, w_{ij}\}$.

Therefore, the clustering problem can be formulated as an optimization problem [24], which is described by Eq. (3):

$$\boldsymbol{P} : \text{minimize}\, z(W, M) = \sum_{i=1}^{n}\sum_{j=1}^{k} w_{ij} d\left(x_i, \mu_j\right) \tag{3}$$

$$\text{subject to} \sum_{j=1}^{k} w_{ij} = 1, \text{ for } i = 1, \ldots, n,$$

$$w_{ij} = 0 \text{ or } 1, \text{ for } i = 1, \ldots, n, \text{ and } j = 1, \ldots, k,$$

where $w_{ij} = 1$ implies object x_i belongs to cluster $G(j)$ and $d(x_i, \mu_j)$ denotes the Euclidean distance between x_i and μ_j for $i = 1, \ldots, n$ and $j = 1, \ldots, k$.

The standard version of the *K*-means algorithm consists of four steps, as shown in **Figure 1**.

The pseudocode of the standard *K*-means algorithm is shown in Algorithm 1.

Algorithm 1. Standard *K*-means algorithm

 # Initialization:
$N: = \{x_1, \ldots, x_n\};$
$M: = \{\mu_1, \ldots, \mu_k\};$
 # Classification:
For $x_i \in N$ **and** $\mu_k \in M$
Calculate the Euclidean distance from each x_i to the k centroids;

Assign object x_i to the closest centroid μ_k;
 # Centroid calculation:
Calculate centroid μ_k;
 # Convergence:
If $M: = \{\mu_1, \ldots, \mu_k\}$ remains unchanged in two consecutive iterations
then:
Stop the algorithm;
else:
Go to **Classification**
End

Since the pioneering studies conducted by Steinhaus [11], Lloyd [12], MacQueen [6], and Jancey [19], many investigations have been aimed at finding a k-partition of N that solves problem $\boldsymbol{P}$, defined by Eq. (3).

It has been shown that the clustering problem belongs to the NP-hard class for $k \geq 2$ or $d \geq 2$ [25, 26]. Therefore, obtaining an optimal solution for an instance of moderate size is generally an intractable problem. Consequently, a variety of heuristic algorithms have been proposed for obtaining the closest possible solution to the optimum of $\boldsymbol{P}$, being the most important of those designed as K-means-type algorithms [6].

It is important to emphasize that the establishment of useful gaps between the optimal solution of the problem $\boldsymbol{P}$ and the solution achieved by K-means remains an open research problem.

The computational complexity of K-means is $O(nkdr)$, where r represents the number of iterations [8, 9], which restricts its use for large instances, because each iteration involves the calculation of all the object-centroid distances. For reducing the complexity of K-means, numerous investigations have been carried out using different strategies for reducing the computational cost and minimizing the objective function.

3. Classification of articles on *K*-means improvements according to the algorithm steps

This section presents a classification of the most relevant articles on improvements to K-means regarding the algorithm steps. The articles were selected applying the systematic review methodology.

3.1 Systematic review process

The search for articles was carried out using four highly prestigious databases: Springer Link, ACM, IEEE Xplore, and ScienceDirect. Queries were issued to these databases using the following search string:

(((({k means} OR {kmeans} OR {Lloyd algorithm} OR {k+means} OR {"k means"} OR {"algoritmo de lloyd"}) AND ({improvement} OR {enhancement} OR {mejora})) AND ({Initialization} OR {inicializacion} OR {beginning} OR {inicio} OR {partition} OR {particion} OR {first step} OR {primer paso} OR {centroide}) AND ({classification} OR {clasificacion} OR {sorting} OR {assignment} OR {asignacion} OR {range search} OR {neighnour search} OR {búsqueda en vecindario}) AND (OR {centroide calculation} OR {calculo de centroide}) AND ({Convergence} OR {Convergencia} OR {Stop criteria} OR {criterio de paro} OR {Stop condition} OR {Condicion de parada} OR {convergence condition} OR {Condicion de convergencia} OR {final step} OR {Paso final})).

As a result of the queries, 1125 articles were retrieved related to the *K*-means algorithm and its improvements. Next, inclusion and exclusion criteria were applied, which reduced the number of articles to 79. The remaining articles were classified according to the algorithm steps as shown in the following subsections.

The flow chart in **Figure 2** shows the steps of the process carried out for selecting the articles. In step "a," the database queries were issued, and a document list was generated, and in step "b," duplicate articles were identified and eliminated. In step "c," based on article titles, those irrelevant to this research were identified and discarded. In step "d," article abstracts were analyzed, and those with little affinity to the subject of study were excluded. In step "e," those documents written in languages different from English or Spanish were eliminated. In step "f," those articles that did not describe an improvement process were discarded. In step "g," the text of the articles was reviewed, and those with little affinity to the subject of study were excluded. In step "h," four articles were eliminated because of possible plagiarism. Finally, in step "i," articles with a small number of citations were discarded; specifically, those with citations below a threshold adjusted by year and category.

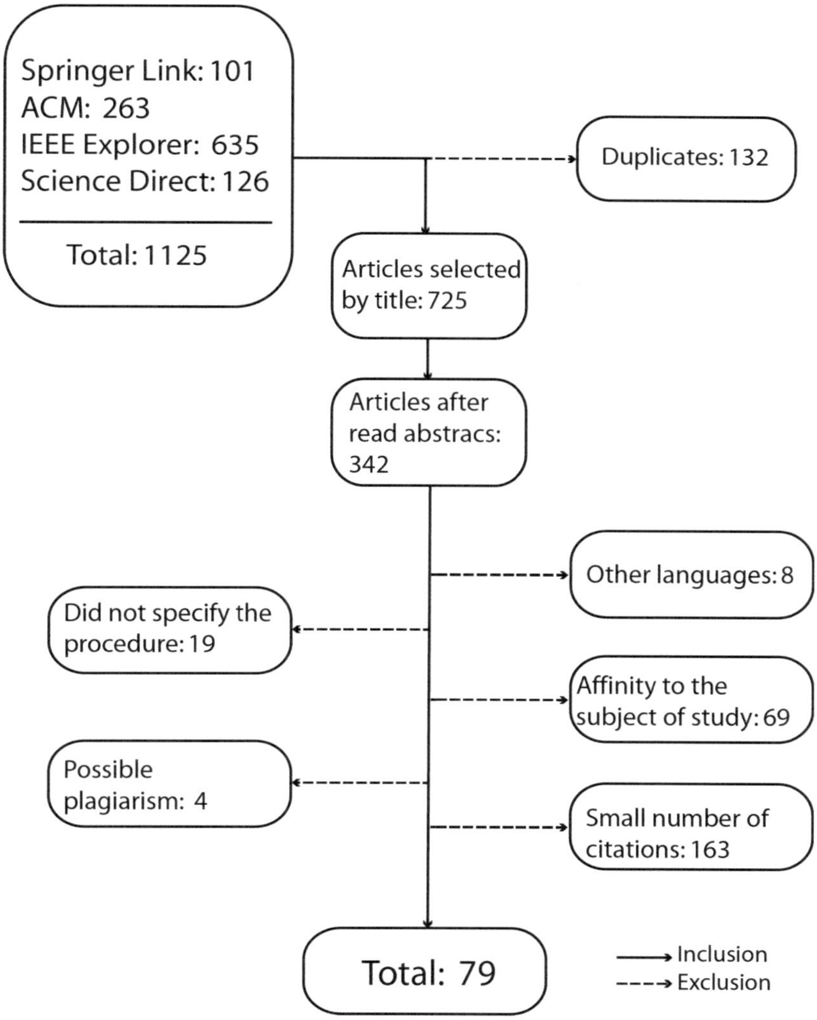

Figure 2.
Process for selecting articles.

3.2 Article classification

As a result of the analysis of the articles, those addressing an improvement to one of the algorithm steps were identified. However, several works were found that involved improvements to more than one step; therefore, the following groups or categories were defined: (a) initialization, (b) classification and centroid calculation, (c) convergence, (d) convergence and initialization, and (e) convergence and classification.

In **Figure 3**, the number of articles for each of the aforementioned groups is shown. Notice that the step with the most articles is initialization and the step with the least attention by researchers is convergence. In the following subsections, the most important articles in each group are briefly described.

3.3 Initialization

The initialization step has received the most attention by researchers, because the algorithm is sensitive to the initial position of the centroids; i. e., different initial centroids may yield different resulting clusters. Consequently, a good initial selection might find a better solution and a reduction in the number of iterations needed by the algorithm to converge.

For this step 38 documents were found about improvements proposed for generating better initial centroids. **Table 1** summarizes information on the articles for this step. Column 1 shows the articles for this step. Columns 2 through 5 indicate the different strategies that researchers have used for obtaining improvements for this step. Finally, column 6 shows the number of articles for each of the strategies.

The second row shows articles on approaches that perform a preprocessing for generating the initial centroids by using particular algorithms or methods. The third row includes articles on methods based on information on data set. The fourth row shows articles on techniques that involve more effective data structures. Finally, the fifth row includes articles where the improvements use other strategies.

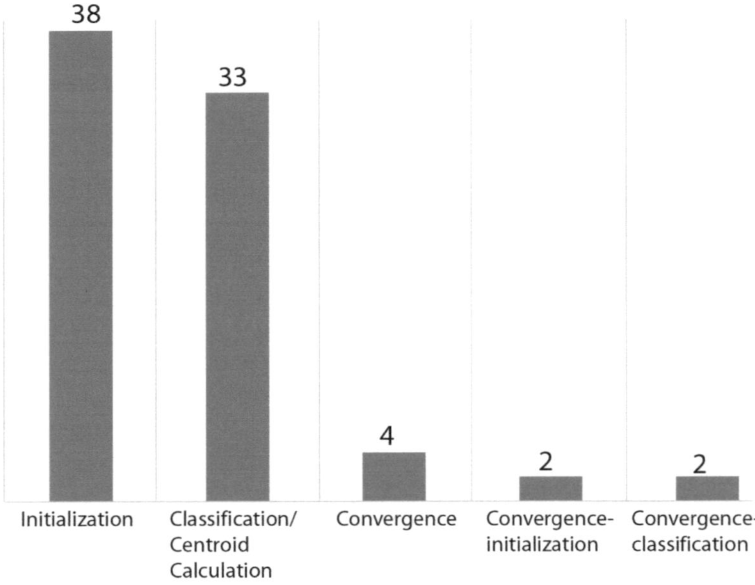

Figure 3.
Number of articles for each of the aforementioned groups.

Articles	Strategy				Number of articles
	Algorithm/ method	Instance information	Data structure	Other	
[24, 27–40]	●				15 (39.47%)
[41–55]		●			15 (39.47%)
[56–59]			●		4 (10.53%)
[60–63]				●	4 (10.53%)

Table 1.
Summary of information on the articles for initialization.

In the rest of this subsection, some of the most important works on the initialization step are summarized. Several of these works mentioned can be used in other algorithms similar to K-means for selecting the initial centroids.

In [27] a clustering method is proposed, where the centroids of the final clusters are used as initial centroids for K-means. The main idea is to randomly select an object x, which is used as a first initial centroid; from this object, the following $k-1$ centroids are selected considering a distance threshold set by the user.

In [28] a modification to Lloyd's work [12] is developed in the field of quantization. The main idea is that objects that are farther from each other have a larger probability of belonging to different clusters; therefore, the strategy proposed consists in choosing an object with the largest Euclidean distance to the rest of the objects for being the first centroid. The following i-th centroids (i = 2, 3, ..., k) will be selected in decreasing order with respect to the largest distance to the first centroid.

In [29] two initialization methods are developed, which are aimed at being applied to large data sets. The proposed methods are based on the densities of the data space; specifically, they need first to divide uniformly the data space into M disjoint hypercubes and to randomly select kN_m/N objects in hypercube m (m = 1, 2, ..., M) for obtaining a total of k centroids.

In [30] a preprocessing is performed called *refining*. This method consists in using K-means for solving M samples of the original data set. The results of SSQ (sum of squared distances) are compared for each of the M solutions, and from the solution with the smallest value, the set of final centroids is extracted, which are used as the initial centroids for solving the entire instance using K-means.

In [31] a preprocessing method is proposed which uses a selection model based on statistics. In particular, it uses the Akaike Information Criterion (AIC) and the Bayesian Information Criterion (BIC) for selecting the set of objects that will be used as initial centroids.

In [42] an algorithm is presented based on two main observations, which state that the more two objects are similar to each other, the largest the possibility that they end up assigned to the same cluster, so they are discarded from the selection of initial centroids. This method is based on density-based multiscale data condensation (DBMSDC) and allows identifying regions with large data densities, and afterwards a list is generated sorting the values by density. Next, select an object according to the sorted list as the first initial centroid, and all the objects that have a ratio inversely proportional to the density of the selected object are discarded. Afterwards, the second centroid is selected as the next object in the list that has not been eliminated, and its surrounding objects are excluded. This process is repeated until all the initial centroids needed are obtained.

A variance-based method for finding the initial centroids is proposed in [44]. First, the method calculates the variances of the values for each dimension, and it selects the dimension with the largest variance. Next, it sorts the objects by the values on the dimension with the largest variance. Finally, it creates k groups with the same number of objects each, and for each group it calculates the median. The medians constitute the initial centroids.

Other researchers have focused their works on using information on data set, such as the distribution of objects and statistical information of them, among others.

In [45] a method is proposed for randomly generating the initial centroids as described next: the first centroid is randomly generated using a uniform probability distribution for the objects; subsequent centroids ($i = 2, 3, ..., k$) are generated calculating a probability that is proportional to the square of their minimal distances to the set of previously selected centroids ($1, ..., i–1$).

In [50] a method is proposed, which is based on a sample of the data set for which an average is calculated. Next, the objects whose distance is larger than the average are identified, and a distance-between-objects criterion is applied for selecting the objects that will constitute the initial objects. The authors claim that this method obtains good results regarding time and solution quality when solving large data sets.

In [55] a method is proposed for eliminating those objects that may cause noise, as well as *outliers*. The method determines the most dense region of the data space, from which it locates the best initial centroids.

By using particular data structures, in [59] a method is presented for estimating the data density in different locations of the space by using kd-tree type structures.

Other researchers [58, 60] have used a combination of genetic algorithms and K-means for the initialization step; however, this method has high computational complexity, since it is necessary to execute the K-means algorithm on the entire set of objects of the population, for each of the generations of the genetic algorithm.

3.4 Classification

Of the four steps of the algorithm, classification is the most time-consuming, because for each object it is necessary to calculate the distance from each object to each centroid.

The 33 articles that are the best proposals for this step are classified in **Table 2**. Column 1 shows the articles related to this step. Columns 2 through 6 indicate the different strategies that researchers have used for achieving improvements for this step. Finally, column 7 shows the number of articles for each of these strategies aiming at reducing the number of calculations of object-centroid distances.

The second row shows articles on approaches that use compression thresholds. The third row includes articles on methods that use information from the initialization step. The fourth row shows articles on techniques that involve more efficient data structures. The fifth row includes articles that present mathematical/statistical processes. Finally, the sixth row shows articles where the improvements use other strategies.

In [64] an improvement is proposed, which reduces the number of calculations of object-centroid distances. For this purpose, an exclusion criterion is defined based on the information of object-centroid distances in two successive iterations: i and $i + 1$. This criterion allows to exclude an object x from the distance calculations to the rest of the centroids, if it is satisfied that the distance to the centroid in iteration $i + 1$ is smaller than that of iteration i.

In [69] a heuristic is proposed, which reduces the number of objects considered in the calculations of object-centroid distances; i.e., the objects with small probability of changing cluster membership are excluded. The rationale behind this heuristic derives from the observation that objects closest to a centroid have a small probability

Articles	Strategy					Number of articles
	Compression thresholds	Information of previous steps	Data structure	Mathematical/ statistical process	Other	
[64–72]	●					9 (27.27%)
[73–80]		●				8 (24.24%)
[81–87]			●			7 (21.21%)
[88–92]				●		5 (15.15%)
[93–96]					●	4 (12.12%)

Table 2.
Summary of information on the articles for classification.

of changing cluster membership, whereas those closer to the cluster border have higher probability of migrating to another cluster. The heuristics determine a threshold for deciding which objects should be excluded. The calculation of the threshold is defined as the sum of the two largest centroid shifts with respect to the previous iteration. Another work with a similar strategy is presented in [66].

In [72] an improvement is presented, which allows excluding from the calculation of object-centroid distances those objects in clusters that have not had object migrations in two successive iterations. This type of clusters is called *stable*, and the objects in such clusters keep their membership unchanged for the rest of the iterations of the algorithm. In the article [73], a similar strategy is presented.

In [74] an improvement to *fast global K-means* algorithm is proposed, which is based on the cluster membership and the geometrical information of the objects. This work also includes a set of inequalities that are used to determine the initial centroids.

In [79] a heuristic is presented, which reduces the number of calculations of object-centroid distances. Specifically, it calculates the distances for each object only to those centroids of clusters that are neighbors of the cluster where the object belongs. This heuristic is based on the observation that objects can only migrate to neighboring clusters.

One of the most representative works for this step is presented in [82], where an improvement is proposed, called *filtering algorithm (FA)*, which uses data structures of binary tree type, called *kd-tree*. Each node of the tree is associated to a set of objects called *cell*. An improvement of this work is described in [85], where the authors claim that it reduces execution time by 33.6% with respect to algorithm FA. Another remarkable improvement to the work in [82] is presented in [83].

3.5 Centroid calculation

Centroid calculation was defined as another step in this analysis, because there exist two variants for this step that differentiate two types of the *K*-means algorithm. In one of the types, centroid calculations are performed once all the objects have been assigned to one cluster. This type of calculation method is called *batch* and is used by [12, 19], among others. The second type of calculation is performed each time an object changes cluster membership. This type of calculation is called *linear K-means* and was proposed by MacQueen. No documents were retrieved related to this step.

3.6 Convergence

The convergence step of the algorithm has received little attention by researchers, which is manifested by the small number of papers on this subject. It is worth mentioning that in recent years, research on this step has produced very promising results concerning the reduction of algorithm complexity at the expense of a minimal reduction of solution quality.

A pioneering work for this step was presented in [97]. The main contribution consisted in associating the values of the squared errors to the stop criterion of the algorithm. In particular, the proposed condition for stopping is when, in two successive iterations, the value of the squared error in one iteration is larger than in the previous one, which guarantees that the algorithm stops at the first local optimum.

Other articles for this step are [98, 99], from the point of view of mathematical analysis, aiming at proving when does the solution obtained reach a global optimum.

3.7 Convergence and initialization

This subsection summarizes two works on improvements for the convergence and initialization steps.

In [100] the stop criterion is associated to the number k of clusters; i.e., in each iteration a new initial centroid is generated for creating a new cluster. This stop criterion is called *incremental*.

In [101] convergence is reached in two ways. In the first condition, the algorithm stops when it reaches a predefined number of iterations. In the second, the algorithm stops when there is no region with a density value larger than a predefined threshold. It is important to mention that in each iteration, the algorithm creates a new cluster guided by the density value in a region.

3.8 Convergence and classification

In this subsection two works are summarized, which present improvements for the convergence and classification steps.

In [102] a stop criterion is proposed, which stops the algorithm when, in ten consecutive iterations, the difference of the squared errors, between iterations i and $i + 1$, does not exceed a predefined threshold.

The work presented in [103] proposes an optimization by integrating the core (classification) of the K-means algorithm and multiple kernel learning using support-vector machines (LS-SVM). By using the Rayleigh coefficient, it optimizes the separation among each group. This algorithm reaches local convergence by obtaining the maximal separation among each of the centroids.

4. Trends

Preceding sections include articles published from the origins of the algorithm up to 2016. This section includes three types of articles: recently published articles on important improvements to K-means, articles that propose improvements to the algorithm implementation using parallel and distributed computing, and articles for new applications of the algorithm.

Regarding improvements to the steps of K-means, several of recently published articles are summarized next. In [104] two algorithm improvements are proposed: one deals with the *outliers* and *noise data* problems, and the other deals with the selection of initial centroids. In [105] two problems are dealt with: the selection of initial centroids and the determination of the number k of clusters. In [106] a new measure of distance or similarity between objects is proposed. In [107] an improvement to the work in [69] is proposed, by defining a new criterion for reducing the processing time of the assignment of objects to clusters. This approach is particularly useful for large instances. In [108] a new stop criterion is proposed, which reduces the number of iterations. In [109] an improvement is proposed for the convergence step of the algorithm aimed at solving large instances. The improvement consists of a new criterion that balances processing time and the quality of the solution. The main idea is to stop the algorithm when the number of objects that change membership is smaller than a threshold.

Recently, in the specialized literature, the parallelization of K-means has been proposed by using MapReduce paradigm [110, 111], which makes possible to process efficiently large instances. In [112] a method is proposed for parallelizing the *K-means++* algorithm [45], which has shown good results for obtaining the initial centroids.

In recent years, a trend has been observed for modifying K-means oriented to new applications, in particular, its application to natural language and text processing. In this regard, one of the remarkable works is presented in [113], in which a modification to K-means is proposed for grouping bibliographic citations. Later, in [114] an improvement is proposed to the algorithm in [113], in order to solve recommendation problems. In [115, 116] an improvement to K-means is proposed for the field of natural language.

5. Conclusions

This chapter presents three aspects of the K-means algorithm: (a) the works that originated the family of K-means algorithms, (b) a systematic review on the algorithm improvements up to 2016, and (c) some of the most recent publications that describe the prospective uses of the algorithm.

Regarding the origin of K-means, it is worth mentioning that it is not only an algorithm but a family of algorithms with the same purpose, which were developed independently in the decades of the 1950s and 1960s.

The systematic review process involved accessing four large databases, from which 1125 documents were retrieved. After applying inclusion and exclusion criteria, 79 documents remained.

Next, we will mention the most important observations organized by subjects:

1. Initialization. Of the four steps of the algorithm, initialization is the step on which the largest number of investigations has focused. The reason for this interest is that the algorithm is highly sensitive to the initial positions of the centroids. Some of the most cited publications are [45, 59].

2. Classification. Most of the works related to this step aim at reducing the number of calculations of object-centroid distances by applying heuristic methods. Some of the most cited works are [73, 82]. Some promising and recently published articles are [72, 107].

3. Centroid calculation. No documents were retrieved related to this step.

4. Convergence. It is remarkable that for this step the number of articles is very small. However, some articles recently published present very promising results by reducing the algorithm complexity without decreasing significantly the solution quality.

Regarding the most recent publications on K-means, improvements have been proposed for solving large instances, as well as parallel and distributed implementations and applications of K-means to new fields such as natural language and text processing, among others.

Finally, because the nature of data clustering is exploratory and applicable to data from many disciplines, it is foreseeable that K-means will continue to be widely used, mainly because of the easiness for interpreting its results.

Author details

Joaquín Pérez-Ortega[1*], Nelva Nely Almanza-Ortega[1], Andrea Vega-Villalobos[1], Rodolfo Pazos-Rangel[2], Crispín Zavala-Díaz[3] and Alicia Martínez-Rebollar[1]

1 National Institute of Technology of Mexico/CENIDET, Cuernavaca, Mexico

2 National Institute of Technology of Mexico/ITCM, Ciudad Madero, Mexico

3 Autonomous University of the State of Morelos, Cuernavaca, Mexico

*Address all correspondence to: jpo_cenidet@yahoo.com.mx

References

[1] Kambalta K, Kollias G, Kumar V, Grama A. Trends in big data analytics. Journal of Parallel and Distributed Computing. 2014;**74**(7):2561-2573. DOI: 10.1016/j.jpdc.2014.01.003

[2] Laney D. 3D Data Management: Controlling Data Volume, Velocity And Variety [Internet]. 2001. Available from: http://blogs.gartner.com/doug-laney/files/2012/01/ad949-3D-Data-Management-Controlling-Data-Volume-Velocity-and-Variety.pdf [Accessed: August 24, 2017]

[3] Raykov YP, Boukouvalas A, Baig F, Little MA. What to do when k-means clustering fails: A simple yet principled alternative algorithm. PLoS One. 2016; **11**(9):e0162259. DOI: 10.1371/journal.pone.0162259

[4] Chun-wei T, Ching-Feng L, Han-Chieh C, Vasilakos AV. Big data analytics: A survey. Journal of Big Data. 2015;**2**(1):21. DOI: 10.1186/s40537-015-0030-3

[5] Wu X, Kumar V, Quinlan JR, Ghosh J, Yang Q, Motoda H, et al. Top 10 algorithms in data mining. Knowledge and Information Systems. 2008;**14**:1-37. DOI: 10.1007/s10115-007-0114-2

[6] MacQueen J. Some methods for classification and analysis of multivariate observations. In: Proc. 5th Berkeley Symp. Math. Statistics and Probability. Vol. 1. 1967. pp. 281-297

[7] Steinley D. K-means clustering: A half-century synthesis. The British Journal of Mathematical and Statistical Psychology. 2006;**59**:1-34

[8] Bock H-H. Origins and extensions of the K-means algorithm in cluster analysis. Journal Electronique d'Histoire des Probabilités et de la Statistique. 2008;**4**(2):1-18

[9] Jain AK. Data clustering: 50 years beyond k-means. Pattern Recognition Letters. 2010;**31**:651-666. DOI: 10.1016/j.patrec.2009.09.011

[10] Blomer J, Lammersen C, Schmidt M, Sohler C. Theoretical analysis of the k-means algorithm—A survey. Algorithm Engineering. 2016;**9220**:81-116. DOI: 10.1007/978-3-319-49487-6_3

[11] Steinhaus H. Sur la division des corps matériels en parties. Bulletin de l'Académie Polonaise des Sciences. Classe 3. 1956;**12**:801-804

[12] Lloyd SP. Least squares quantization in PCM. In: Bell Telephone Labs Memorandum, Murray Hill NJ. Reprinted in: IEEE Trans. Information Theory IT-28. Vol. 2. 1982. pp. 129-137

[13] Diday E. Une nouvelle méthode de classification automatique et reconnaissance des formes: la méthode des nuées dynamiques. Revue de Statistique Appliquée. 1971;**XIX**:19-33

[14] Diday E. The dynamic clusters method in nonhierarchical clustering. International Journal of Computing and Information Sciences. 1973;**2**:61-88

[15] Diday E, Govaert G. Classification avec distance adaptive. Comptes Rendus de l'Académie des Sciences. 1974;**278A**: 993-995

[16] Bock H-H. Automatische Klassifikation: Theoretische und Praktische Methoden zur Strukturierung von Daten (Clusteranalyse). Göttingen: Vandenhoeck & Ruprecht; 1974

[17] Anderberg MR. Cluster Analysis for Applications. New York: Academic Press; 1973

[18] Späth H. Cluster Analyse Algorithmen zur Objektklassifizierung

und Datenreduktion. Oldenbourg Verlag, München K Wien. English Translation: Cluster Analysis Algorithms for Data Reduction and Classification of Objects. Chichester, UK: Ellis Horwood Ltd; 1980

[19] Jancey RC. Multidimensional group analysis. Australian Journal of Botany. 1966;**14**:127-130

[20] Forgy EW. Cluster analysis of multivariate data: Efficiency versus interpretability of classifications. In: Biometric Society Meeting, Riverside, California. Abstract in Biometrics. Vol. 21. 1965. p. 768

[21] Pearson K. On the coefficient of racial likeness. Biometvika. 1926;**18**: 105-117

[22] Rao CR. The use of multiple measurements in problems of biological classification. Journal of the Royal Statistical Society: Series B: Methodological. 1948;**10**(2):159-203

[23] Sokal RR. Distance as a measure of taxonomic similarity. Systematic Zoology. 1961;**10**(2):70-79

[24] Selim SZ, Ismail MA. K-means-type algorithms: A generalized convergence theorem and characterization of local optimality. IEEE Transactions on Pattern Analysis and Machine Intelligence. 1984;**1**:81-87

[25] Aloise D, Deshpande A, Hansen P, Popat P. NP-hardness of Euclidean sum-of-squares clustering. Machine Learning. 2009;**75**(2):245-248

[26] Mahajan M, Nimbhorkar P, Varadarajan K. The planar k-means problem is NP-hard. Theoretical Computer Science. 2012;**442**:13-21

[27] Tou JT, Gonzalez RC. Pattern Recognition Principles. USA: Addison-Wesley; 1974

[28] Katsavounidis I, Kou J, Zhang Z. A new initialization technique for generalized Lloyd iteration. IEEE Signal Processing Letters. 1994;**1**:144-146

[29] Moh'd B, Roberts SA. New methods for the initialization of clusters. Pattern Recognition Letters. 1996;**17**:451-455. DOI: 10.1016/0167-8655(95)00119-0

[30] Bradley PS, Fayyad UM. Refining initial points for K-means clustering. In: Proceeding of the 15th International Conference on Machine Learning (ICML98). San Francisco: Morgan Kaufmann; 1998. pp. 91-99

[31] Pelleg D, Moore A. X-means: Extending K-means with efficient estimation of the number of clusters. In: Proceedings of the Seventieth International Conference on Machine Learning (ICML); July 2000; Palo Alto, CA

[32] Su T, Dy JG. In search of deterministic methods for initializing k-means and gaussian mixture clustering. Intelligent Data Analysis. 2007;**11**: 319-338. DOI: 10.3233/IDA-2007-11402

[33] Zalik KR. An efficient k'-means clustering algorithm. Pattern Recognition Letters. 2008;**29**:1385-1391. DOI: 10.1016/j.patrec.2008.02.014

[34] Nazeer KA, Sebastian MP. Improving the accuracy and efficiency of the k-means clustering algorithm. In: Proceedings of the World Congress on Engineering; 1-3 July 2009; London, UK. 2009. pp. 1-3

[35] Lee D, Baek S, Sung K. Modified k-means algorithm for vector quantizer design. IEEE Signal Processing Letters. 1997;**4**(1):2-4. DOI: 10.1109/97.551685

[36] Ahmed AH, Ashour W. An initialization method for the k-means algorithm using RNN and coupling degree. International Journal of Computer Applications. 2011;**25**:1-6

[37] Nazeer KA, Kumar SD, Sebastian MP. Enhancing the k-means clustering algorithm by using a O(n logn) heuristic method for finding better initial centroids. In: International Conference on Emerging Applications of Information Technology; 19-20 February 2011; Kolkata, India. 2011. pp. 261-264

[38] Salaman R, Kecman V, Li Q, Strack R, Test E. Two stage clustering with k-means algorithm. Recent Trends in Wireless and Mobile Networks. 2011; **162**:110-122. DOI: 10.1007/978-3-642-21937-5_11

[39] Celebi ME. Improving the performance of k-means for color quantization. Image and Vision Computing. 2011;**29**:260-271. DOI: 10.1016/j.imavis.2010.10.002

[40] Zhanguo X, Shiyu C, Wentao Z. An improved semi-supervised clustering algorithm based on initial center points. Journal of Convergence Information Technology. 2012;**7**:317-324

[41] Yuan F, Meng ZH, Zhang HX, Dong CR. A new algorithm to get the initial centroids. In: Proceeding of 2004 International Conference on Machine Learning and Cybernetics; 26-29 August 2004; Shanghai. China: IEEE; 2004. pp. 1191-1193

[42] Khan SS, Ahmad A. Cluster center initialization algorithm for K-means clustering. Pattern Recognition Letters. 2004;**25**:1293-1302. DOI: 10.1016/j.patrec.2004.04.007

[43] Pham DT, Dimov SS, Nguyen CD. Selection of k in k-means clustering. Proceedings of the Institution of Mechanical Engineers-Part C: Journal of Mechanical Engineering Science. 2005; **219**:103-119. DOI: 10.1243/095440605X8298

[44] Al-Daoud. A new algorithm for cluster initialization. In: The Second World of Enformatika Conference. 2005

[45] Arthur D, Vassilvitskii S. K-means++: The advantages of careful seeding. In: Proceedings of the Eighteenth Annual ACM-SIAM Symposium on Discrete Algorithms. 2007. pp. 1027-1035

[46] Zhang Z, Zhang J, Xue H. Improved k-means clustering algorithm. Signal & Image Processing. 2008;**4**:169-172

[47] Vanisri D, Loganathan D. An efficient fuzzy clustering algorithm based on modified k-means. International Journal of Engineering, Science and Technology. 2010;**2**:5949-5958

[48] Yedla M, Pathakota SR, Srinivasa TM. Enhancing k-means clustering algorithm with improved initial center. International Journal of Computer Science & Information Technology. 2010;**2**:121-125

[49] Xie J, Jiang S. A simple and fast algorithm for global k-means clustering. In: 2010 Second International Workshop on Education Technology and Computer Science; 6-7 March 2010. Wuhan, China: IEEE; 2010. pp. 36-40

[50] Eltibi M, Ashour W. Initializing k-means algorithm using statistical information. International Journal of Computer Applications. 2011;**29**:51-55

[51] Li CS. Cluster center initialization method for k-means algorithm over data sets with two clusters. Procedia Engineering. 2011;**24**:324-328. DOI: 10.1016/j.proeng.2011.11.2650

[52] Xie J, Jiang S, Xie W, Gao X. An efficient global k-means clustering algorithm. Journal of Computers. 2011; **6**:271-279

[53] Elagha M, Ashour WM. Efficient and fast initialization algorithm for k-means clustering. International Journal of Intelligent Systems and Applications. 2012;**4**:21-31

[54] Zhang Y, Cheng E. An optimized method for selection of the initial centers of K-means clustering. Lecture Notes in Computer Science. 2013;**8032**: 149-156. DOI: 10.1007/978-3-642-39515-4_13

[55] Abudaker M, Ashour W. Efficient data clustering algorithms: Improvements over K-means. International Journal of Intelligent Systems and Applications. 2013;**3**:37-49. DOI: 10.5815/ijisa.2013.03.04

[56] Alsabti K, Ranka S, Singh V. An efficient k-means clustering algorithm. In: Electrical Engineering and Computer Science. 1997. p. 43

[57] Li M, Ng MK, Cheung YM, Huang JZ. Agglomerative fuzzy k-means clustering algorithm with selection of number of clusters. IEEE Transactions on Knowledge and Data Engineering. 2008;**20**(11):1519-1534. DOI: 10.1109/TKDE.2008.88

[58] Laszlo M, Mukherjee S. A genetic algorithm using hyper-quadtrees for low dimensional k-means clustering. IEEE Transactions on Pattern Analysis and Machine Intelligence. 2006;**28**:533-543. DOI: 10.1109/TPAMI.2006.66

[59] Redmond SJ, Heneghan C. A method for initializing the K-means clustering algorithm using kd-trees. Pattern Recognition Letters. 2007;**28**: 965-973. DOI: 10.1016/j.patrec.2007.01.001

[60] Babu GP, Murty MN. A near-optimal initial seed value selection in k-means algorithm using a genetic algorithm. Pattern Recognition Letters. 1993;**14**:763-769. DOI: 10.1016/0167-8655(93)90058-L

[61] Steinley D. Local optima in K-means clustering: What you don't know may hurt you. Psychological Methods. 2003; **8**(3):294-304. DOI: 10.1037/1082-989X.8.3.294

[62] Tian J, Lin Z, Suqin Z, Lu L. Improvement and parallelism of k-means clustering algorithm. Tsinghua Science and Technology. 2005;**10**: 277-281. DOI: 10.1016/S1007-0214(05)70069-9

[63] Hand JD, Krzanowski WJ. Optimising k-means clustering results with standard software packages. Computational Statistics and Data Analysis. 2005;**49**:969-973. DOI: 10.1016/j.csda.2004.06.017

[64] Fahim AM, Salem AM, Torkey FA, Ramadan MA. An efficient enhanced k-means clustering algorithm. Journal of Zhejiang University. 2006;**7**:1626-1633. DOI: 10.1631/jzus.2006.A1626

[65] Tsai C, Yang C, Chiang M. A time efficient pattern reduction algorithm for k-means based clustering. In: IEEE International Conference on Systems, Man and Cybernetics; 1-10 October 2007. Montreal, Quebec, Canada: IEEE; 2008. pp. 504-509

[66] Chiang M, Tsai C, Yang C. A time-efficient pattern reduction algorithm for k-means clustering. Information Sciences. 2011;**181**:716-731. DOI: 10.1016/j.ins.2010.10.008

[67] Singh RV, Bhatia MP. Data clustering with modified k-means algorithm. In: 2011 International Conference on Recent Trends in Information Technology (ICRTIT). 3-5 June 2011. Chennai, India: IEEE; 2011. pp. 717-721

[68] Lee S, Lin J. An accelerated k-means clustering algorithm using selection and erasure rules. Journal of Zhejiang University. Science. 2012;**13**:761-768. DOI: 10.1631/jzus.C1200078

[69] Perez J, Pires CE, Balby L, Mexicano A, Hidalgo M. Early classification: A new heuristic to improve the classification step of k-means. Journal of

Information and Data Management. 2013;**4**:94-103

[70] Yusoff IA, Isa NAM, Hasikin K. Automated two-dimensional k-means clustering algorithm for unsupervised image segmentation. Computers and Electrical Engineering. 2013;**39**:907-917. DOI: 10.1016/j.compeleceng.2012.11.013

[71] Mexicano A, Rodriguez R, Cervantes S, Ponce R, Bernal W. Fast means: Enhancing the k-means algorithm by accelerating its early classification version. AIP Conference Proceedings. 2015;**1648**:820004-1-820004-4. DOI: 10.1063/1.4913023

[72] Perez J, Pazos R, Hidalgo M, Almanza N, Diaz-Parra O, Santaolaya R, et al. An improvement to the k-means algorithm oriented to big data. AIP Conference Proceedings. 2015;1648: 820002-1-820002-4. DOI: 10.1063/1.4913021

[73] Lai JZC, Huang T, Liaw Y. A fast k-means clustering algorithm using cluster center displacement. Pattern Recognition. 2009;**42**:2551-2556. DOI: 10.1016/j.patcog.2009.02.014

[74] Lai JZC, Huang T. Fast global k-means clustering using cluster membership and inequality. Pattern Recognition. 2010;**43**:1954-1963. DOI: 10.1016/j.patcog.2009.11.021

[75] Al-Zoubi M, Hudaib A, Hammo B. New efficient strategy to accelerate k-means clustering algorithm. American Journal of Applied Sciences. 2008;**5**: 1247-1250

[76] Chang C, Lai JZC, Jeng M. A fuzzy k-means clustering algorithm using cluster center displacement. Journal of Information Science and Engineering. 2011;**27**:995-1009

[77] Bagirov AM, Ugon J, Webb D. Fast modified global k-means algorithm for incremental cluster construction. Pattern Recognition. 2011;**44**:866-876. DOI: 10.1016/j.patcog.2010.10.018

[78] Osamor VC, Adebiyi EF, Oyelade JO, Doumbia S. Reducing the time requirement of k-means algorithm. PLoS One. 2012;**7**:1-10. DOI: 10.1371/journal.pone.0049946

[79] Perez J, Mexicano A, Santaolaya R, Hidalgo M, Moreno A, Pazos R. Improvement to the K-means algorithm through a heuristic based on a bee honeycomb structure. In: Fourth World Congress on Nature and Biologically Inspired Computing; 5-9 November 2012; Mexico. Mexico: IEEE; 2013. pp. 175-180

[80] Bai L, Liang J, Siu C, Dang C. Fast global k-means clustering based on local geometrical information. Information Sciences. 2013;**245**:168-180. DOI: 10.1016/j.ins.2013.05.023

[81] Phillips SJ. Acceleration of k-means and related clustering algorithms. Lecture Notes in Computer Science. 2002;**2409**:166-177. DOI: 10.1007/3-540-45643-0_13

[82] Kanungo T, Mount DM, Netanyahu NS, Piatko CD, Silverman R, Wu AY. An efficient k-means clustering algorithm: Analysis and implementation. IEEE Transactions on Pattern Analysis and Machine Intelligence. 2002;**24**:881-892

[83] Vrahatis MN, Boutsinas B, Alevizos P, Pavlides G. The new K-windows algorithm for improving the K-means clustering algorithm. Journal of Complexity. 2002;**18**:375-391. DOI: 10.1006/jcom.2001.0633

[84] Napoleon D, Lakshmi PG. An efficient k-means clustering algorithm for reducing time complexity using uniform distribution data points. In: Trendz in Information Sciences and Computing (TISC2010), 17-19

December 2010. Chennai, India: IEEE; 2011. pp. 42-45

[85] Lai JZC, Liaw Y. Improvement of the k-means clustering filtering algorithm. Pattern Recognition. 2008; **41**:3677-3681. DOI: 10.1016/j.patcog.2008.06.005

[86] Hamerly G, Drake J. Accelerating Lloyd's algorithm for k-means clustering. In: Cebeli M, editor. Partitional Clustering Algorithms. Springer: Cham; 2015. pp. 41-78. DOI: 10.1007/978-3-319-09259-1_2

[87] Wang J, Wang J, Ke Q, Zeng G, Shipeng L. Fast approximate k-means via cluster closures. In: Multimedia Data Mining and Analytics. 2015. Springer International Publishing AG. Cham. pp. 373-395. DOI: 10.1007/978-3-319-14998-1_17

[88] Cofarelli C, Nieddu L, Seref O, Pardalos PM. K-T.R.A.C.E: A kernel k-means procedure for classification. Computers and Operations Research. 2007;**34**:3154-3161. DOI: 10.1016/j.cor.2005.11.023

[89] Salaman R, Kecman V, Li Q, Strack R, Test E. Fast k-means algorithm clustering. International Journal of Computer Networks and Communications. 2011;**3**. DOI: 10.5121/ijcnc.2011.3402

[90] Kaur N, Sahiwal JK, Kaur N. Efficient k-means clustering algorithm using ranking method in data mining. International Journal of Advanced Research in Computer Engineering & Technology. 2012;**1**: 85-91

[91] Scitovski R, Sabo K. Analysis of the K-means algorithm in the case of data points occurring on the border of two or more clusters. Knowledge-Based Systems. 2014;**57**:1-7. DOI: 10.1016/j.knosys.2013.11.010

[92] Xu L, Hu Q, Hung E, Szeto C. A heuristic approach to effective an efficient clustering on uncertain objects. Knowledge-Based Systems. 2014;**66**: 112-125. DOI: 10.1016/j.knosys.2014.04.027

[93] Elkan C. Using the triangle inequality to accelerate k-means. In: Proceedings of the Twentieth International Conference on Machine Learning (ICML-2013); Washington, DC. 2003

[94] Fahim AM, Saake G, Salem AM, Torkey FA, Ramadan MA. K-means for spherical clusters with large variance in sizes. International Journal of Scientific Research and Innovative Technology. 2008;**2**:2923-2928

[95] Sarma TH, Viswanath P, Reddy BE. A hybrid approach to speed-up the k-means clustering method. International Journal of Machine Learning and Cybernetics. 2013;**4**(2):107-117. DOI: 10.1007/s13042-012-0079-7

[96] Pakhira MK. A modified k-means algorithm to avoid empty clusters. International Journal of Recent Trends in Engineering. 2009;**1**:220-226

[97] Perez J, Pazos R, Cruz L, Reyes G, Basave R, Fraire H. Improving the efficiency and efficacy of the K-means clustering algorithm through a new convergence condition. In: International Conference on Computational Science and its Applications (ICCSA 2007). 2007

[98] Samma A, Salam R. Adaptation of k-means algorithm for image segmentation. World Academy of Science, Engineering and Technology. 2009;**50**:58-62

[99] Bottou L, Bengio Y. Convergence properties of the K-means algorithms. In: Advances in Neural Information Processing Systems 7, Tesauro G,

Touretzky D, editors. Cambridge, MA: The MIT Press; 1995:586-592

[100] Pham DT, Dimov SS, Nguyen CD. An incremental K-means algorithm. Proceedings of the Institution of Mechanical Engineers, Part C: Journal of Mechanical Engineering Science. 2004; **218**:783-795. DOI: 10.1243/0954406041319509

[101] Likas A, Vlassis N, Verbeek JJ. The global K-means clustering algorithm. Pattern Recognition. 2003;**36**:451-461. DOI: 10.1016/S0031-3203(02)00060-2

[102] Lam YK, Tsang PWM. eXploratory K-means: A new simple and efficient algorithm for gene clustering. Applied Soft Computing. 2012;**12**:1149-1157. DOI: 10.1016/j.asoc.2011.11.008

[103] Yu S, Tranchevent L, Liu X, Glanzel W, Suykens JAK, Moor B, et al. Optimized data fusion for kernel k-means clustering. IEEE Transactions on Pattern Analysis and Machine Intelligence. 2012;**34**:1031-1039. DOI: 10.1109/TPAMI.2011.255

[104] Yu S, Chu S, Wang C, Chan Y, Chang T. Two improved k-means algorithms. Applied Soft Computing. 2018;**68**:747-755. DOI: 10.1016/j.asoc.2017.08.032

[105] Zhang G, Zhang C, Zhang H. Improved K-means algorithm based on density canopy. Knowledge-Based Systems. 2018;**145**:289-297. DOI: 10.1016/j.knosys.2018.01.031

[106] Wang R, Li H, Chen M, Dai Z, Zhu M. MIC-KMeans: A maximum information coefficient based high-dimensional clustering algorithm. Advances in Intelligent Systems and Computing. 2019;**764**:208-218. DOI: 10.1007/978-3-319-91189-2_21

[107] Perez J, Almanza N, Ruiz J, Pazos R, Saenz S, Lelis J, et al. A-means: Improving the cluster assignment phase of k-means for big data. International Journal of Combinatorial Optimization Problems and Informatics. 2018;**9**(2):3-10

[108] Mexicano A, Rodriguez R, Cervantes S, Montes P, Jimenez M, Almanza N, et al. The early stop heuristic: A new convergence criterion for k-means. In: AIP Conference Proceedings 2016. AIP Publishing; 2016. p. 310003

[109] Perez J, Almanza N, Romero D. Balancing effort and benefit of K-means clustering algorithms in big data realms. PLoS One. 2018;**13**(9):1-19. DOI: 10.1371/journal.pone.0201874

[110] Zhao W, Ma H, He Q. Parallel k-means clustering based on MapReduce. Lecture Notes in Computer Science. 2009;**5931**:674-679. DOI: 10.1007/978-3-642-10665-1_71

[111] Moertini VS, Venica L. Enhancing parallel k-means using MapReduce for discovering knowledge from big data. In: IEEE International Conference on Cloud Computing and Big Data Analysis (ICCCBDA); 5-7 July 2016. Chengdu, China: IEEE; 2016. pp. 81-87

[112] Bahmani B, Moseley B, Vattani A, Kumar R, Vassilvitskii S. Scalable K-means ++. Proceedings of the VLDB Endowment. 2012;**5**:622-633. DOI: 10.14778/2180912.2180915

[113] McCallum A, Nigam K, Ungar L. Efficient clustering on high-dimensional data sets with application of reference matching. In: International Conference on Knowledge Discovery and Data Mining; 20-23 August 2000. Boston, Massachusetts: ACM; 2000. pp. 169-178

[114] Li J, Zhang K, Yang X, Wei P, Wang J, Mitra K, et al. Category preferred canopy-k-means based collaborative filtering algorithm. Future Generation Computer Systems. 2019;**93**: 1046-1054. DOI: 10.1016/j.future.2018.04.025

[115] Hussain S, Haris M. A k-means based co-clustering (KCC) algorithm for sparse, high dimensional data. Expert Systems with Applications. 2019;**118**: 20-34. DOI: 10.1016/j.eswa.2018.09.006

[116] Naeem S, Wumaier A. Study and implementing K-mean clustering algorithm on English text and techniques to find the optimal value of K. International Journal of Computer Applications. 2018;**182**(31):7-14. DOI: 10.5120/ijca2018918234

Chapter 6

"Set of Strings" Framework for Big Data Modeling

Igor Sheremet

Abstract

The most complicated task for big data modeling in comparison with relational approach is its variety, being a consequence of heterogeneity of sources of data, accumulated in the integrated storage space. "Set of Strings" Framework (SSF) provides unified solution of this task by representation of database as updated finite set of facts, being strings, in which structure is defined by current metadatabase, which is also an updated set of the context-free generating rules. This chapter is dedicated to SSF formal and substantial description.

Keywords: big data, "Set of Strings" framework, string databases, context-free grammars, Post systems

1. Introduction

From the three "V's," traditionally used for description of big data (volume, variety, velocity) [1–7], variety is the most difficult for theoretical modeling. The main reason of such difficulty is heterogeneity of sources of data, accumulated in the integrated storage space. By this, data items, passing to the aforementioned storage, have different structures and formats (more or less formalized texts, multimedia, hyperlinked trees of pages, etc.), which makes practically impossible application to such data of well-known relational-originated approaches to database (DB) description, manipulation, and knowledge extraction/application [8–12]. This obstacle makes hardly achieved the fourth "V" (veracity), which last time is often associated with big data [13–17], as well as with implementation of data mining over such data storages [18–21].

Such background makes necessary the alternative approach to data and knowledge modeling. This chapter contains compact consideration of the so-called "Set of Strings" Framework (SSF), developed in order to integrate on the unified theoretical basis capabilities, already used in the relational-like data representations and associated with them knowledge models, with big data immanent property—its variety.

SSF is a result of an attempt to design the aforementioned basis upon the most general representation of elementary data item, which may be stored, transported, received, processed, and visualized. Such representation is string (no matter, symbol, or bit), and SSF combines the best features of classical string-generating formal grammars, developed by Chomsky [22], with string-operating logical systems, proposed by Post [23].

The second section of this chapter is dedicated to the description of string databases (SDB), while the third, - to their interconnections with relational and non-relational DB. In the fourth section, incomplete information modeling within SSF is considered. The main content of the fifth section are the so-called word equations on context-free languages (WECFL), being key element of the SSF algorithmics.

2. "Set of strings" basic equations

The background of the SSF is representation of a database as a finite set of strings:

$$W_t = \{w_1, ..., w_{m(t)}\} \subset V^*, \tag{1}$$

where W_t means DB at the discrete time moment t and V^* is a set of all strings in the initial (terminal) alphabet V. Such databases will be called lower, if it is necessary to distinguish them from the other, the *set of strings databases*. The structure of DB elements $w_i \in W_t$, named *facts*, is determined by metadatabase (MDB), in which the current state is denoted by D_t.

Couple

$$\Theta_t = \langle W_t, D_t \rangle, \tag{2}$$

is named data storage (DS). Data storage is in the correct state, if $W_t \in \mathbf{W}(D_t)$, where $\mathbf{W}(D_t)$ is the set of all correct databases, defined by the MDB.

Access message to DS is triple:

$$\omega_t = \langle o, c, x \rangle, \tag{3}$$

where o is the operation, which execution is the purpose of the access (insert, delete, update, query), c is the DS component (DB, MDB) which is the objective of the access, and x is the content of the access, i.e., query body, or DB elements (facts), which are inserted or deleted. For simplicity it is supposed that the answer (reply) to the access is obtained by the user at the moment $t+1$, next to t, and it is denoted A_{t+1}, if c = DB, and A^D_{t+1}, if c = MDB (both sets are finite).

A set of all possible access messages (3) is called data storage manipulation language (DSML).

SSF background is a sequential definition of four interconnected representations of DSML semantics.

Set-theoretical (S)-semantics of DSML is defined by equations on sets, which connect together input data, DB before and after access, and answer (reply) to the access.

Mathematical (M)-semantics follows aforementioned equations but is defined by some well-known and understandable mathematical constructions, being background of DSML.

Operational (O)-semantics is adequate to M-semantics but is represented by algorithms, providing execution of operations on DB.

At last, implementational (I)-semantics is also represented by algorithms, which, in general case, are much more efficient than the previous, in which the main purpose is recognition of algorithmic decidability of answer search (derivation), i.e., possibility of answer generation by finite number of steps.

Let us begin from **S-semantics** of the DSML segment, addressing DB, called lower, as usually, data manipulation language (DML). The equations, defining DML S-semantics, operate the following sets:

1. W_t (database at the moment of user's access to DB).

2. W_{t+1} (database after execution of operation, i.e., at the moment $t+1$, when answer is accepted by the user).

3. I_t (set, expressing user's awareness about some fragment of problem area at the moment of access to DB).

4. A_{t+1} (answer to the access).

Basic equations, defining DML S-semantics, are as follows:

$$W_{t+1} = W_t \cup I_t, \tag{4}$$

$$A_{t+1} = W_{t+1} - W_t, \tag{5}$$

for insertion (speaking more precisely, inclusion),

$$W_{t+1} = W_t - I_t, \tag{6}$$

$$A_{t+1} = W_t - W_{t+1}, \tag{7}$$

for deletion (exclusion),

$$W_{t+1} = W_t, \tag{8}$$

$$A_{t+1} = W_t \cap I_t, \tag{9}$$

and for query (everywhere "$-$" is subtraction on sets). As seen, Eqs. (4)–(9) fully correspond to the sense of basic operations on DB, inherent to any DML. In Eqs. (6) and (9), set I_t may be infinite.

Example 1. Let database, containing data items from various emergency devices, be as follows:

$$\begin{aligned} W_t = \{ & \mathit{AREA\ GREEN\ VALLEY\ IS\ IN\ NORMAL\ STATE\ AT}\ 15.03, \\ & \mathit{AREA\ BLUE\ LAKE\ IS\ IN\ NORMAL\ STATE\ AT}\ 15.05, \\ & \mathit{AREA\ LOWER\ FOREST\ IS\ SMOKED\ AT}\ 15.20\}, \end{aligned} \tag{10}$$

(due to free use of natural language in facts, it is unnecessary to comment DB content). Equation

$$W_{t+1} = W_t \cup \{\mathit{AREA\ GREEN\ VALLEY\ IS\ SMOKED\ AT}\ 15.20\} \tag{11}$$

describes insertion of data item, in which the source is device, mounted at the Green Valley, which was detected as smoked since 15.20. When at this moment $t+1$ user accesses DB with query, in which the purpose is to get information about all smoked areas, the infinite set I_t may be as follows:

$$\begin{aligned} I_{t+1} = \{ & \mathit{AREA\ A\ IS\ SMOKED\ AT}\ 00.00, \ldots, \\ & \mathit{AREA\ A\ IS\ SMOKED\ AT}\ 23.59, \ldots, \\ & \mathit{AREA\ AA\ IS\ SMOKED\ AT}\ 00.00, \ldots, \\ & \mathit{AREA\ AA\ IS\ SMOKED\ AT}\ 23.59, \ldots, \\ & \mathit{AREA\ Z\ IS\ SMOKED\ AT}\ 00.00, \ldots, \\ & \mathit{AREA\ Z\ IS\ SMOKED\ AT}\ 23.59, \ldots\}. \end{aligned} \tag{12}$$

The answer to the query is

$$A_{t+2} = W_{t+1} \cap I_{t+1} = \{AREA\ GREEN\ VALLEY\ IS\ SMOKED\ AT\ 15.20,\ AREA\ LOWER\ FOREST\ IS\ SMOKED\ AT\ 15.20\}. \quad (13)$$

In expression (12), names of all areas are strings in the alphabet $V = \{A, \ldots, Z, 0, \ldots, 9, .,\}$, so

$$I_{t+1} = \{AREA\} \bullet V^* \bullet \{SMOKED\ AT\} \bullet \{00, \ldots, 23\} \bullet \{.\} \bullet \{00, \ldots, 59\}. \blacksquare \quad (14)$$

Note that definitions (4)–(9) are not unique. For example, in the inclusion definition, elements of I_t set, having place in the DB at moment t, may be included to the answer

$$A_{t+1} = W_t \cap I_t, \quad (15)$$

as well as the answer may be defined as

$$A_{t+1} = \{FACT''\} \bullet (W_t \cap I_t) \bullet \{''ALREADY\ PRESENTS\ IN\ DATABASE\} \cup \{FACT''\} \bullet (W_{t+1} - W_t) \bullet \{''IS\ INCLUDED\ TO\ DATABASE\}. \quad (16)$$

So, according to Eq. (16), the answer to the access may be as follows:

$$A_{t+1} = \{FACT\ \text{“}AREA\ GREEN\ VALLEY\ SMOKED\ AT\ 15.20\text{”}ALREADY\ PRESENTS\ IN\ DATABASE,\ FACT\ \text{“}AREA\ LOWER\ FOREST\ IS\ SMOKED\ AT\ 15.20\text{”}\ IS\ INCLUDED\ TO\ DATABASE\}. \blacksquare \quad (17)$$

As may be seen, Eqs. (4)–(9) are based on the closed-world interpretation, which defines that the absence of the fact in the database is equivalent to its absence in the real world (problem area).

DML operations do not touch MDB; thus $D_{t+1} = D_t$.

Let us consider DML **M- and O-semantics** of DML.

The background of M-semantics of the simplest DML is the representation of the MDB D_t as a set of the context-free (CF) generating rules $\alpha \to \beta$, where α is a nonterminal symbol (“nonterminal” for short) and β is a string of both nonterminal and terminal symbols. Every nonterminal symbol, from the substantial point of view, is the name of some substring of fact, entering DB; thus β represents the structure of α. The only nonterminal symbol α_0, which does not enter any string β, is the “axiom” in the terminology of formal grammars and “fact” in the terminology of SSF. So MDB D_t unambiguously defines CF grammar

$$G_t = \langle V, N_t, \alpha_0, D_t \rangle, \quad (18)$$

where

$$N_t = \{\alpha \mid \alpha \to \beta \in D_t\} \quad (19)$$

is the set of nonterminals (“nonterminal alphabet”) of G_t.

Database W_t is named *correct to metadatabase* D_t, if

$$W_t \subseteq L(G_t), \quad (20)$$

i.e., facts, having place in the DB, are words of the CF language $L(G_t)$. In other notation,

$$(\forall w \in W_t)\ \alpha_0 \underset{G_t}{\overset{*}{\Longrightarrow}} w, \tag{21}$$

where $\underset{G_t}{\overset{*}{\Longrightarrow}}$ is used to define that string in alphabet $V \cup N_t$ is generated (or derived) from another one.

Example 2. Let MDB D_t be as follows (nonterminal symbols are framed by metalinguistic brackets):

$$
\begin{array}{l}
<fact> \rightarrow AREA <name\ of\ area> IS <state> \\
\qquad AT <time>, \\
<name\ of\ area> \rightarrow <text>, \\
<state> \rightarrow IN\ NORMAL\ STATE, \\
<state> \rightarrow SMOKED, \\
<time> \rightarrow <hours>.<minutes>, \\
<hours> \rightarrow <0\ to\ 1><0\ to\ 9>, \\
<hours> \rightarrow 2<0\ to\ 3>, \\
<0\ to\ 1> \rightarrow 0, \\
<0\ to\ 1> \rightarrow 1, \\
<0\ to\ 9> \rightarrow 0, \\
\dots \\
<0\ to\ 9> \rightarrow 9, \\
<0\ to\ 3> \rightarrow 0, \\
\dots \\
<0\ to\ 3> \rightarrow 3, \\
<minutes> \rightarrow <0\ to\ 5><0\ to\ 9>, \\
<0\ to\ 5> \rightarrow 0, \\
\dots \\
<0\ to\ 5> \rightarrow 5, \\
<text> \rightarrow <symbol>, \\
<text> \rightarrow <symbol><text>, \\
<symbol> \rightarrow A, \\
\dots \\
<symbol> \rightarrow Z, \\
<symbol> \rightarrow 0, \\
\dots \\
<symbol> \rightarrow 9, \\
<symbol> \rightarrow \sqcup.
\end{array}
$$

Database

$$W_t = \{\ AREA\ AW\ IS\ SMOKED\ AT\ 15.10,\quad AREA\ E\ IS\ IN\ NORMAL\ STATE\ AT\ 23.59\ \}$$

is correct to this MDB, unlike database
$W_t = \{AREA\ AT\ NORMAL\}$.∎

Proposed application of CF grammars differs from the classical, in which the main sense is the description of a set of correct sentences of some language (most frequently, programming language). This description is created by its developers or researchers, is based on syntactic categories referred as nonterminals, and is constant through all life cycle of the language (minor changes may be done by reason of language modification or deeper understanding). In the SSF case, CF generating rules are used for description of the DB element (facts) structure, so nonterminals are more semantic than syntactic objects. From the other side, MDB is updated by DS administration and is a dynamic set, in which changes provide immediate changes of DB in order to keep it in the correct state. Such changes may be defined by the following equations, similar to Eqs. (4)–(9):

$$D_{t+1} = D_t \cup I_t^D, \tag{22}$$

$$A_{t+1}^D = D_{t+1} - D_t, \tag{23}$$

$$W_{t+1} = W_t \tag{24}$$

for insertion (inclusion) of new CF rules to MDB,

$$D_{t+1} = D_t - I_t^D, \tag{25}$$

$$A_{t+1}^D = D_t - D_{t+1}, \tag{26}$$

$$W_{t+1} = W_t \cap L(G_{t+1}) \tag{27}$$

for deletion (exclusion) of CF rules, having place in MDB,

$$D_{t+1} = D_t, \tag{28}$$

$$A_{t+1}^D = D_t \cap I_t^D, \tag{29}$$

$$W_{t+1} = W_t \tag{30}$$

and for query to MDB. Here I_t^D is similar to I_t in Eqs. (4)–(9), being a set of CF rules representing knowledge of DS administration about MDB. As seen, Eqs. (22) and (23) provide extension of MDB; thus

$$L(G_t) \subseteq L(G_{t+1}), \tag{31}$$

and DB remains correct, because

$$W_t \subseteq L(G_t) \subseteq L(G_{t+1}). \tag{32}$$

In Eqs. (25) and (26), where some part (subset) of MDB may be deleted,

$$L(G_{t+1}) \subseteq L(G_t), \tag{33}$$

so some facts $w \in L(G_t)$ may become not satisfying condition (20) of DB correctness to MDB D_{t+1}, because $w \notin L(G_{t+1})$. In Eqs. (25)–(30), it is presumed, that D_t is also SDB, in which MDB defines structure of CF rules, which may be as Example 2.

Let us note that the notion of SDB correctness to MDB is from the substantial point of view weaker than the notion of *data storage correctness*, because in general case

$$\boldsymbol{W}(D_t) \subseteq 2^{L(G_t)}, \tag{34}$$

i.e., set of databases in correct storage is the subset of Boolean of $L(G_t)$, while SDB correct to MDB is such that

$$\boldsymbol{W}(D_t) = 2^{L(G_t)}, \tag{35}$$

i.e., every SDB, containing facts, being words of CF language $L(G_t)$, is correct, which is not true in the reality. DS correctness is the generalization of notion of DB integrity, deeply developed inside relational approach covering the total content of database, i.e., interconnections between its different elements. There are known various tools for integrity criteria declaration and check—first of all, functional dependencies and their multiple modifications [24–32]. Storage correctness, being SDB analog of integrity, is considered inside SSF on the basis of augmented Post systems (APS).

Let us consider now the application of the described segment of the SSF to the representation of the most frequently used data models. We shall call such application by the term "emulation."

3. Emulation of the known data models

We shall demonstrate how relational and non-relational databases may be represented on the described higher background. We shall consider relational data model as a full-scope example of databases with symmetric access (DBSA) [8–12] and a family of asymmetric access (or key-addressed) databases (KADB), which contains, among others, hypertext, page, and WWW- and Twitter-like DB [32–41].

Let us begin from the **relational model of data**.

Consider relational database (RDB), in which the scheme is $\{R_1(A_1^1, \ldots, A_{m1}^1), \ldots, R_k(A_1^k, \ldots, A_{mk}^k)\}$, where $R_1, \ldots, R_k$ are the names of relations and $A_1^1, \ldots, A_j^i, \ldots, A_{mk}^k$ are the names of attributes. Every relation R_i at moment t is the set of tuples

$$R_i \subseteq D_j^i \times \ldots \times D_{mj}^i, \tag{36}$$

where D_j^i is the domain (set of possible values of attribute A_j^i).

We shall define SDB $<W_t, D_t>$, corresponding to this RDB, as follows. We shall include to the MDB D_t rules

$$\begin{aligned} &<fact> \rightarrow R_1 : <A_1^1>, \ldots, <A_{m1}^1>, \\ &\qquad \ldots \\ &<fact> \rightarrow R_k : <A_1^k>, \ldots, <A_{mk}^k>, \end{aligned} \tag{37}$$

where "<" and ">" are the dividers (aforementioned metalinguistic brackets in the Backus-Naur notation) and R_i and A_j^i are the strings, being names of relations and attributes, respectively (dividers provide syntactic unambiguity).

Along with Eq. (36), MDB will include rules such that

$$D_j^i = \left\{ b | < A_j^i > \overset{*}{\Rightarrow} b \& b \in \{V - \{,\}\}^* \right\}, \tag{38}$$

i.e., these rules provide generation of sets of words in terminal alphabet V, being domains of the respective attributes. For unambiguity we assume that comma "," does not enter values of attributes $V \in D_j^i$.

By this, every tuple $(b_1^i, \ldots, b_{mi}^i)$ of the relation R_i is represented by fact

$$R_i : b_1^i, \ldots, b_{mi}^i \in W_t. \tag{39}$$

Note that representation of facts in the form (37)–(39) is not unique. As seen from Examples 1 and 2, tuples, entering relations, may be represented as any natural language phrases, described by the corresponding rules.

Let us consider now **key-addressed databases**. Their common feature is that every fact, entering KADB, includes a unique key, which is necessary to select, delete, and update this fact. These DB are associated with NoSQL family of DML [42–45], which in the last years is considered as a practical alternative to SQL-like DML [8–12, 46–48], developed since the introduction of the relational model of data.

We shall represent KADB as set

$$W_t = \{k_1 = d_1, \ldots, k_m = d_m\}, \tag{40}$$

where symbol "=" inside angle brackets is the divider, $k_i \in (V - \{=\})^*$ is the key, and $d_i \in V^*$ is the data, corresponding to this key (or identified by it). At every moment t, KADB must satisfy the consistency condition: KADB is consistent, if

$$(\forall t)(\forall k \in (V - \{=\})^*) |\{k =\} \cdot V^* \cap W_t| \leq 1, \tag{41}$$

i.e., set W_t would not include two or more elements with one and the same key.

Content of access to KADB must include key k, so $I_t \subseteq \{k =\} \cdot V^*$,

and for inclusion $I_t = \{k = d\}$. S-semantics of insertion to KADB is as follows:

$$W_{t+1} = \begin{cases} W_t \cup \{k = d\}, if \{k =\} \cdot V^* \cap W_t = \{\varnothing\}, \\ W_t - \{k =\} \cdot V^* \cup \{k = d\} otherwise, \end{cases} \tag{42}$$

because postulation of fact $k = d$ at moment t is equivalent to the negation of fact $k = d'$, where $d \neq d'$, which was postulated at some earlier moment $t' < t$. So in the case of KADB update is implemented by insertion, and reply to this access may be defined as follows:

$$A_{t+1} = \begin{cases} k = d, if \{k =\} \cdot V^* \cap W_t = \{\varnothing,\}, \\ (W_t \cap \{k =\} \cdot V^*) \cup \{k = d\} otherwise, \end{cases} \tag{43}$$

thus in the case of update reply contains deleted as well as included fact.

Concerning M-semantics of KADB, we may see, that every known class of such databases is identified by its own structure of keys and techniques of their extraction from the current processed fact.

The simplest approach is implemented in Twitter network, where keys, necessary for access to the descendants of the current element of the hypertext, are bounded by two dividers—"#" from the left and blank from the right.

In the Internet HTTP/WWW service, similar keys are represented as strings of symbols, visualized by other colors in comparison with the rest of the text of the current hypertext page. This is equivalent to splitting terminal alphabet V to two subsets, first including symbols of the ordinary colors and the second symbols of the "key-representing" colors. However, HTTP/WWW hypertexts are organized in a much more complicated manner. First of all, along with the displayed pages, in which the structure is described by hypertext markup language (HTML) or its various later versions (XML et al.), there is another KADB, in which elements contain keys, being the aforementioned strings of another colors, and data are, in fact, unified resource locators (URL), providing direct network access to the subordinated pages. This access is possible, because URL contains string, providing application of the domain name service (DNS) for resolving proper IP address. In fact, HTML is no more than language for the convenient representation of CF rules, which form current metadatabase of the WWW KADB.

One of the simplest versions of KADB is the so-called page databases, in which elements are strings of equal length, the first string of the page being key [38, 39]. Thus

$$(\forall t)L(G_t) = V^l\left(V^l\right)^*, \tag{44}$$

where l is the length of the string (in this case divider "=" is redundant). Data may be also string $p : d$, where ":" is the divider and prefix p before the sequence of l-symbol strings defines the name of the program, called for this sequence interpretation (e.g., visualization). In general case d may be the string of bits, not only string of symbols of alphabet V.

Until now we discussed only S-semantics and start point of M-semantics, being representation of metadatabase as a set of rules of CF grammar. Second such point in the SSF is the representation of databases with incomplete information.

4. Representation of databases with incomplete information and sentential data manipulation languages

Let D_t be the metadatabase. Then database with incomplete information (for short, DBI or, if it is necessary to underline "set of strings" DBI, then SDBI), denoted X_t, is the finite set of the so-called incomplete facts (N-facts) being sentential forms (SF) of CF grammar G_t:

$$X_t = \{x_1, \dots, x_m\} \subseteq SF(G_t), \tag{45}$$

where

$$SF(G_t) = \left\{ x | a_0 \overset{*}{\underset{G_t}{\Rightarrow}} x \right\} \tag{46}$$

is the set of all sentential forms of grammar G_t. Obviously, $L(G_t) \subset SF(G_t)$.

Example 3. Consider MDB from Example 2 and corresponding DBI

$$\begin{aligned} X_t = \{&AREA\ LONELY\ TREES\ IS\ NORMAL\ AT\ 12.31, \\ &AREA\ LONELY\ TREES < state > AT\ 12. < minutes >, \\ &AREA < name\ of\ area > IS\ SMOKED\ AT\ 15.30\}. \end{aligned}$$

The first N-fact of the three, entering this DBI, does not contain nonterminals, so it is fact in the sense of S-semantics of DML. The second N-fact contains nonterminals <*state*> and <*minutes*>, which correspond to the uncertainty of the state of the area Lonely Trees and time moment, when this state occurs; however, the aforementioned moment enters interval from 12.01 to 12.59. The last N-fact contains information about the same area, which was detected as smoked at 15.30. ■

Before consideration of equations, defining M-semantics of operations on DBI, we shall introduce interpretation of relation $\overset{*}{\underset{G_t}{\Longrightarrow}}$ of the mutual derivability of sentential forms of context-free grammar as relation of mutual informativity of N-facts.

Let G_t be acyclic and unambiguous CF grammar [49]. If so, $\overset{*}{\underset{G_t}{\Longrightarrow}}$ is the relation of partial order on the set $SF(G_t)$ [38, 39].

There is maximal element of the set $SF(G_t)$—it is axiom α_0 ("fact") because for every $x \in SF(G_t)$, $\alpha_0 \overset{*}{\underset{G_t}{\Longrightarrow}} x$. For every subset $X \subseteq SF(G_t)$, there exists set of its upper bounds, e.g., sentential forms ("N-facts") $y \in SF(G_t)$, such that $y \overset{*}{\underset{G_t}{\Longrightarrow}} x$ for all $x \in X$, and minimal (least) upper bound, sup X, such that for every other upper bound y from the mentioned set, the relation $y \overset{*}{\underset{G_t}{\Longrightarrow}} \sup X$ is true. For some $X \subseteq SF(G_t)$, there may exist set of lower bounds, e.g., sentential forms ("N-facts") $y \in SF(G_t)$, such that $x \overset{*}{\underset{G_t}{\Longrightarrow}} y$ for all $x \in X$, and maximal lower bound inf X such that for every other lower bound y, $\inf X \overset{*}{\underset{G_t}{\Longrightarrow}} y$ is true.

Algorithms of sup and inf generation are described in detail in [38, 39].

Example 4. For DBI from Example 3, inf X_t does not exist, but

$$\sup X_t = AREA < name\ of\ area > IS < state > AT\ 1 < 0\ to\ 9 > . < minutes > .$$

At the same time,

$$\begin{gathered} \inf \{AREA\ LONELY\ TREES\ IS < state > AT\ 12.30, \\ AREA < name\ of\ area > IS\ SMOKED\ AT\ 12. < minutes > \} \\ = AREA\ LONELY\ TREES\ IS\ SMOKED\ AT\ 12.30. \blacksquare \end{gathered}$$

Since now we shall use interpretation of $\overset{*}{\underset{G_t}{\Longrightarrow}}$ as of the relation of the *mutual informativity* of incomplete facts. According to this interpretation, $x \overset{*}{\underset{G_t}{\Longrightarrow}} x'$ means that N-fact x' is not less informative in comparison with N-fact x (if $x \overset{+}{\underset{G_t}{\Longrightarrow}} x'$, then x' is more informative and is called *concretization of* x). (This interpretation naturally fits to A. Kolmogorov's algorithmic theory of information basic postulates, i.e., constructive objects mutual complexity [50]).

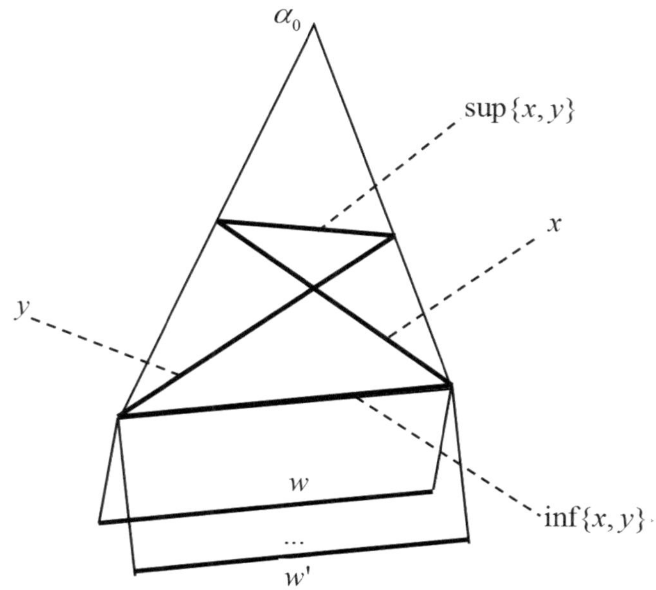

Figure 1.
Graphical illustration of interconnection between $\sup\{x,y\}$ *and* $\inf\{x,y\}$.

Graphical illustration of interconnections between $\sup\{x,y\}$ and $\inf\{x,y\}$ is in **Figure 1**. As seen, $\sup\{x,y\}\overset{*}{\Rightarrow}x$, $\sup\{x,y\}\overset{*}{\Rightarrow}y$, $x\overset{*}{\Rightarrow}\inf\{x,y\}$, $y\overset{*}{\Rightarrow}\inf\{x,y\}$, $\inf\{x,y\}\overset{*}{\Rightarrow}w$, and $\inf\{x,y\}\overset{*}{\Rightarrow}w'$, where $w\in L(G_t)$ and $w'\in L(G_t)$.

Let us consider DBI $X_t=\{x_1,\dots,x_{m(t)}\}\subseteq SF(G_t)$. We shall call such DBI nonredundant (NR), if there are no two N-facts x and x' entering X_t, one of which is more informative than the other. (It is obvious that if $x'\underset{G_t}{\overset{*}{\Longrightarrow}}x$, then there is no necessity of storing x', because all information, having place in x', presents in x. So x' is a redundant N-fact, and DBI, containing such N-facts, is also redundant).

Until the contrary is declared, we shall consider only NR DBI lower. By this, when defining M-semantics of update of NR DBI, understood as inclusion of N-fact $x\in SF(G_t)$, it is reasonable to suppose, that it contains maximally informative N-facts, which only may be acquired by the system. In this case inclusion of N-fact $x\in X_t$ to DBI may be defined as follows:

$$X_{t+1}=X_t\bigcup\{x\}-\left\{y\mid y\in X_t\&\left(x\underset{G_t}{\overset{*}{\Longrightarrow}}y\bigvee y\underset{G_t}{\overset{*}{\Longrightarrow}}x\right)\right\}. \qquad (47)$$

According to this definition, N-fact x inclusion to DBI X_t causes extraction from DBI of all N-facts, which are more or less informative than x. Such logics provides maintenance of nonredundancy of the DBI. As seen, all N-facts, having place in X_t and being "compatible" with N-fact x by informativity, are eliminated from this DBI.

It is reasonable to define reply to the inclusion of N-fact x as set of N-facts, eliminated from DBI:

$$A_{t+1}=X_t-X_{t+1}. \qquad (48)$$

Example 5. If N-fact $x=$ *AREA GREEN VALLEY IS SMOKED AT* 15.30 is included to DBI X_t from Example 3, then

$$X_{t+1} = \{AREA\ LONELY\ TREES\ IS\ NORMAL\ AT\ 12.31,$$
$$AREA\ LONELY\ TREES <state> AT\ 12. <minutes>,$$
$$AREA\ GREEN\ VALLEY\ IS\ SMOKED\ AT\ 15.30\},$$

because

$$AREA <name\ of\ area> SMOKED\ AT\ 15.30 \overset{*}{\underset{G_t}{\Longrightarrow}}$$

$$AREA\ GREEN\ VALLEY\ IS\ SMOKED\ AT\ 15.30. \blacksquare$$

As to M-semantics of queries, there may by two different versions, which run out of the new DBI features in comparison with DB.

The *first version* is obvious:

$$A_{t+1}^{y} = \left\{x \mid x \in X_t \& y \overset{*}{\Rightarrow} x\right\}, \tag{49}$$

where A_{t+1}^{y} is the answer to the query with content y, and all N-facts from DBI X_t, which are no less informative than x, are included to the answer.

As seen, by this we postulate query language to databases (or DB with complete information): it is the set of sentential forms of CF grammar G_t, and in the case of SF y is the content of query ω_t,

$$I_t = \left\{w | y \overset{*}{\underset{G_t}{\Rightarrow}} w \& w \in V^*\right\}, \tag{50}$$

i.e., it is the set of facts, more informative than N-fact y, having place in the query. So combining Eqs. (9) and (49), we obtain

$$A_{t+1} = W_t \cap I_t = \left\{w | y \overset{*}{\underset{G_t}{\Rightarrow}} w \& w \in W_t\right\}, \tag{51}$$

i.e., the result of access is the subset of database W_t, containing all facts, more informative than y.

The background of the *second* version is the interpretation of the query as an action, which aim is to check if there are such possible facts $w \in L(G_t)$, which are more informative than N-fact x, entering X_t, and N-fact y (query content):

$$(\exists w \in L(G_t))\ x \overset{*}{\underset{G_t}{\Longrightarrow}} w \& y \overset{*}{\underset{G_t}{\Longrightarrow}} w, \tag{52}$$

so, while x is not concretization of y, it is sensible to include x to the answer, because there may be facts $w \in L(G_t)$, which are both x and y concretizations. The set of such facts is the intersection $W_x \cap W_y$, where

$$W_x = \left\{w \mid x \overset{*}{\underset{G_t}{\Longrightarrow}} w\right\}, \tag{53}$$

$$W_y = \left\{w \mid y \overset{*}{\underset{G_t}{\Longrightarrow}} w\right\}.$$

Finite representation of the aforementioned intersection is, obviously, maximal lower bound of the set $\{x, y\}$. For this reason, answer to the query with content y may be set

$$\overline{A}_{t+1}^{y} = \{x \mid x \in X_t \ \ \& \ \exists \inf\{x, y\}\}, \tag{54}$$

and, as an alternative,

$$\overline{\overline{A}}_{t+1}^{y} = \{\inf\{x, y\} \mid x \in X_t \ \& \ \exists \ \inf\{x, y\}\}. \tag{55}$$

Example 6. Consider DBI X_t from Example 3 and the query with content $y = AREA < name\ of\ area > IS\ SMOKED\ AT < time >$. The purpose of this query is to get information about all areas smoked. According to Eqs. (49), (54), and (55)

$$A_{t+1}^{y} = \{AREA < name\ of\ area > IS\ SMOKED\ AT\ 14.30\},$$

$$\overline{A}_{t+1}^{y} = \{AREA\ LONELY\ TREES < state > AT\ 13. < minutes >,$$
$$AREA < name\ of\ area > IS\ SMOKED\ AT\ 14.30\},$$

$$\overline{\overline{A}}_{t+1}^{y} = \{AREA\ LONELY\ TREES\ IS\ SMOKED\ AT\ 13. < minutes >,$$

$$AREA < name\ of\ area > IS\ SMOKED\ AT\ 14.30\}. \blacksquare$$

Returning to DB, which DML S-semantics was defined by Eqs. (4)–(9), we may now write its M-semantics equations not only for query but also for insertion and deletion. Namely, if string w is the content of the insertion access, then

$$W_{t+1} = \begin{cases} W_t \cup \{w\}, \ if\, w \in L(G_t), \\ W_t \quad otherwise. \end{cases} \tag{56}$$

Similarly, if string y, containing terminal and nonterminal symbols of CF grammar G_t, is the content of the delete access, then

$$W_{t+1} = W_t - \left\{ w | y \overset{*}{\underset{G_t}{\Rightarrow}} w \& w \in L(G_t) \right\}. \tag{57}$$

Data manipulation languages, described in this section, will be called sentential (SDML), because content of any access to DB/DBI, specified with the help of such DML, is a sentential form of CF grammar, which set of rules is current MDB.

Concerning KADB, capabilities of the sentential DML are compatible with the aforementioned NoSQL languages, which provide selection of DB elements, in which keys are specified in the queries.

Of course, it is not difficult to extend SDML by features, providing construction of more complicated selection criteria (including, e.g., number intervals) [38, 39]. But in comparison with SQL and similar relational languages, providing symmetric access to DB, SDML are rather poor. To achieve capabilities of the relationally complete query languages, it is necessary to extend SDML by features, providing comparison of values, having place in different facts. Such features are critically needed also for knowledge representation, extraction, and processing.

To achieve the formulated purpose, we shall use another tool, differing from CF grammars, namely, Post systems (PS), which also operate strings but, due to variables in their basic constructions (productions), have basic capabilities for aforementioned functions. The result of integration of the described "set of strings" databases with PS are augmented Post systems. The intermediate layer between SDB and APS is formed by the so-called word equations on context-free languages, considered in the next section.

5. Word equations on context-free languages

Word equation is a well-known object of discrete mathematics, defined as follows [51–56].

Word equation is written as

$$s = s', \tag{58}$$

where s and s' are the so-called terms. Term is a non-empty sequence of symbols of alphabet, which we shall call terminal, presuming it is the same set V, as higher, and variables, which universum is denoted Γ. So $s \in (V \cup \Gamma)^+, s' \in (V \cup \Gamma)^+$. *Domain* (set of values) of every variable $\gamma \in \Gamma$, having place in any term, is V^*. Term without any variables is, obviously, word in alphabet V. At least one variable must present in WECFL or just the same in term ss' (or $s's$):

$$ss' \in (V \cup \Gamma)^+ - V^+. \tag{59}$$

Set

$$d = \{\gamma_1 \to u_1, ..., \gamma_n \to u_n\}, \tag{60}$$

where $\gamma_1, ..., \gamma_n$ are the variables, $u_1, ..., u_n$ are the strings in alphabet V, and $\to$ is the divider (which is not occasionally the same as higher in the generating rules $\alpha \to \beta$, entering metadatabases), is called *substitution*.

Term $s[d]$ is the result of application of substitution d to term s and is defined as follows. If

$$s = \overline{u_1}\gamma_{i1}\overline{u_2}...\overline{u_m}\gamma_{im}\overline{u}_{m+1}, \tag{61}$$

where $\overline{u_i} \in V^*$ and $i = 1, ..., m + 1,$, then

$$s[\delta] = \overline{u_1}\,\overline{\gamma_{i1}}\,\overline{u_2}\,...\overline{u_m}\,\overline{\gamma_{im}}\,\overline{u}_{m+1}, \tag{62}$$

where

$$\overline{\gamma_{ij}} = \begin{cases} u_{ij}, \textit{if}\, \gamma_{ij} \to u_{ij} \in d \\ \gamma_{ij} \;\; \textit{otherwise}. \end{cases} \tag{63}$$

Definitions (62) and (63) cover general case, when some of the variables, entering term s, do not enter the substitution (60).

Substitution d is called *terminal substitution* to term $s \in (V \cup \Gamma)^+ - V^+$, if

$$s[d] \in V^+, \tag{64}$$

i.e., result of its application to term is word in the alphabet V. In this case, obviously,

$$\{\gamma_{i1}, ..., \gamma_{in}\} \subseteq \{\gamma_i, ..., \gamma_n\}. \tag{65}$$

Terminal substitution to terms s and s' is called *solution of word equation* (58), if

$$s[d] \equiv s[d'], \tag{66}$$

i.e., result of application of d to terms s and s' is one and the same word (here $'' \equiv ''$ is identity sign).

Returning to SDB and M-semantics of their DML, we may see that *set of terms may be the simplest query language to SDB*. If term

$$s = \overline{u_1}\gamma_{i1}\overline{u_2}...\overline{u_m}\gamma_{im}\overline{u}_{m+1}, \tag{67}$$

is query to DB W_t, then

$$I_t = \{\overline{u_1}u_{i1}\overline{u_2}...\overline{u_m}u_{im}\overline{u}_{m+1}|u_{i1} \in V^* \&...\&u_{im} \in V^*\}, \tag{68}$$

and

$$A_{t+1} = W_t \cap I_t = \{w|w \in W_t\&(\exists u_{i1} \in V^*)...(\exists u_{im} \in V^*)\overline{u_1}u_{i1}\overline{u_2}...\overline{u_m}u_{im}\overline{u}_{m+1} = w\}, \tag{69}$$

so Eq. (69) is the definition of M-semantics of the term's query language to SDB; as seen, $w \in A_{t+1}$, if $w \in W_t$, and word equation $s = w$ has at least one solution.

Example 7. Consider database W_t, containing three facts:
SENSOR 1 IS AT GREEN VALLEY,
SENSOR 2 IS AT BLUE LAKE,
AREA LOWER FOREST IS SMOKED.

If query $s = SENSOR\ a$, which purpose, as seen, is to select all facts with information about sensor installation, then

$$A_{t+1} = \left\{ \begin{array}{l} SENSOR\quad 1\quad IS\quad AT\quad GREEN\ VALLEY, \\ SENSOR\quad 2\quad IS\quad AT\quad BLUE\ LAKE \end{array} \right\},$$

and solution of word equations
SENSOR a = *SENSOR 1 IS AT GREEN VALLEY*
and
SENSOR a = *SENSOR 2 IS AT BLUE LAKE*
are, respectively,

$$\{a \to 1\quad IS\quad AT\quad GREEN\ VALLEY\}$$

and

$$\{a \to 2\quad IS\quad AT\quad BLUE\ LAKE\}. \blacksquare$$

However, the application of the term's query language to databases with incomplete information, containing sentential forms of CF grammar with scheme D_t, being DS metadatabase, is not so simple and needs more sophisticated mathematical background.

Let G be CF grammar, corresponding metadabase D (lower index t for simplicity is omitted). We shall call **word equation on context-free language** $L(G)$ couple

$$<s = s', \delta>, \tag{70}$$

where the first component $s = s'$ is the word equation in the sense (58), called here *kernel*, while

$$\delta = \{\gamma_1 \rightarrow \beta_1, ..., \gamma_l \rightarrow \beta_l\}, \tag{71}$$

is the so-called suffix, which defines domains (sets of values) of variables $\gamma_1, ..., \gamma_l$, entering terms s and s', by means of strings $\beta_1, ..., \beta_l$, containing terminal and nonterminal symbols of grammar G. Kernel and suffix must satisfy the so-called sentential condition

$$\{s[\delta], s'[\delta]\} \subseteq SF(G), \tag{72}$$

i.e., strings, being the result of application of substitution δ to terms s and s', must be sentential forms of grammar G. As seen, δ is the generalization of substitution (60), so it will be called lower *SF-substitution*.

WECFL (70) may be read "$s = s'$, where δ."

Domain of variable γ_i is the set of strings in terminal alphabet V, which are generated from string β_i by application of rules of grammar G. This domain is denoted as

$$\mathrm{V}(\gamma_i, \delta) = \left\{u | \gamma_i \rightarrow \beta_i \in \delta \& \beta_i \overset{*}{\Rightarrow} u \& u \in V^*\right\} \tag{73}$$

(from here we shall use $\overset{*}{\Rightarrow}$ in the sense $\underset{G}{\overset{*}{\Rightarrow}}$).

Suffix δ defines *set of terminal substitutions* to terms s and s', denoted

$$\sum\nolimits_{\delta} = \bigcup_{\substack{u_1 \in V(\gamma_1, \delta) \\ \cdots \\ u_l \in V(\gamma_l, \delta)}} \{\{\gamma_1 \rightarrow u_1, ..., \gamma_l \rightarrow u_l\}\}. \tag{74}$$

As it is easy to see, direct consequence of the sentential condition (72) and definition (74) is

$$\{s[d], s'[d]\} \subseteq L(G), \tag{75}$$

for every $d \in \sum_{\delta}$.

If terminal substitution d is such that

$$s[d] \equiv s[d'], \tag{76}$$

it is called *solution of WECFL* (70). Set of solutions of WECFL (70), which is infinite in general case, is denoted $D[s = s', \delta]$.

Function V may be applied to every term, so

$$\mathrm{V}(s, \delta) = \left\{s[d] | d \in \sum\nolimits_{\delta}\right\} \subseteq L(G), \tag{77}$$

$$\mathrm{V}(s', \delta) = \left\{s'[d] | d \in \sum\nolimits_{\delta}\right\} \subseteq L(G). \tag{78}$$

Example 8. Let metadatabase be the same as in Example 2, and WECFL is

$$<AREA\ GREEN\ VALLEY\ IS a = b\ AT 15.00,$$
$$\{a \rightarrow\ <state>AT<time>,\ b \rightarrow AREA<name\ of\ area>IS<state>\}>.$$

As seen, this equation satisfies sentential condition, because

$$s[\delta] = AREA\ \ GREEN\ \ VALLEY\ IS < state > AT < time > \in SF(G_t),$$
$$s'[\delta] = AREA < name\ of\ area > IS < state > AT\ \ 15.00 \in SF(G_t).$$

According to Eq. (73),
$\mathrm{V}(a,\delta) = \{IN\ NORMAL\ STATE AT\} \cdot T \cup \{SMOKED AT\} \cdot T\mathrm{V}(b,\delta) = \{AREA\}$ $\cdot S \cdot \{IS\ IN\ NORMAL\ STATE AT\} \cup \{AREA\} \cdot S \cdot \{IS\ SMOKED\}$, where T is the set of strings, explicating time (00.00, 00.01, ..., 23.58, 23.59), while S is the set of names of the monitored areas.

Terminal substitution

$$s = a \rightarrow SMOKED AT\, 15.00, \quad b \rightarrow AREA\ \ GREEN\ \ VALLEY\ IS\ SMOKED$$

is the solution of the presented WECFL. ∎

As seen, in general case the set of solutions of WECFL may be infinite, and the problem is to find finite representation of this set.

Let us consider two sentential forms x and x' of unambiguous and acyclic CF grammar G. Each of them defines generated (derived) from it set of strings, being words of language $L(G)$:

$$W_x = \left\{ w | x \overset{*}{\Rightarrow} w \& w \in V^* \right\}, \tag{79}$$

$$W_y = \left\{ w | y \overset{*}{\Rightarrow} w \& w \in V^* \right\}. \tag{80}$$

And therefore SF x and x' are finite representations of sets W_x and W_y, both being subsets of language $L(G)$. This obstacle serves as background for the following statement, representing necessary solution.

Statement 1 [51]. If

$$W = W_x \cap W_{x'} \neq \{\varnothing\}, \tag{81}$$

then there exists SF y such that

$$W = \left\{ w | x \overset{*}{\Rightarrow} y \& x' \overset{*}{\Rightarrow} y \& y \overset{*}{\Rightarrow} w \& w \in V^* \right\}. \blacksquare \tag{82}$$

Verbally, non-empty intersection of sets W_x and W_y is the subset of language $L(G)$, in which words are generated from $SF\ y$, which itself is generated from $SF\ x$ and x' simultaneously.

Example 9. Consider SF $s[d]$ and $s'[d]$ from Example 8. As seen, SF

$$y = AREA\ \ GREEN\ \ VALLEY\ IS < state > AT\ \ 15.00$$

is the finite representation of intersection $W_{s[d]} \cap W_{s'[d]}$.∎

Thus SF y from Eq. (82) is nothing else than required finite representation of the non-empty intersection (81).

This finding is a basis for constructing the set of solutions $D[s = s', \delta]$. Let us begin from the case where all variables, having place in WECFL (or, just the same, in term ss'), enter it once, i.e., there is no more than one occurrence of any variable in ss'.

Obviously, if

$$W = \mathrm{V}(s,\delta) \cap \mathrm{V}(s',\delta) = \{\varnothing\}, \tag{83}$$

then WECFL (70) does not have a solution, i.e.,

$$D[s = s', \delta] = \{\varnothing\}, \tag{84}$$

and if $W \neq \{\varnothing\}$, then, since $s[\delta]$ and $s'[\delta]$ are sentential forms of CF grammar G, there exists finite representation of set W, being SF y generated (derived) from $s[\delta]$ and $s'[\delta]$ simultaneously.

From this place it is clear, that finite representation of the set $D[s = s', \delta]$ is set

$$\overline{\delta} = \{\gamma_1 \to \overline{\beta}_1, \ldots, \gamma_l \to \overline{\beta}_l\}, \tag{85}$$

such that

$$W = \{w | s[\overline{\delta}] = s'[\overline{\delta}] = y \& y \overset{*}{\Rightarrow} w \& w \in V^*\}. \tag{86}$$

It is easy to verify that $\overline{\beta}_1, \ldots, \overline{\beta}_l$ are strings, containing terminal and nonterminal symbols, and being generated from strings $\beta_1, \ldots, \beta_l$, respectively, by $s[\delta] \Rightarrow y^*$ and $s'[y] \Rightarrow y^*$.

Set $\overline{\delta}$ will be named *unifier of WECFL* (70). In accordance with Eqs. (54) and (55), we shall consider lower so-called maximal unifiers, corresponding to

$$y = \inf\{s[\delta], s'[\delta]\}, \tag{87}$$

where y is the maximal lower bound of the considered two-element set.

Example 10. Let metadatabase be the same as in Example 2, and WECFL is

$$< AREA\ a\ IS\ SMOKED\ AT\ t \ = \ b\ AT\ 15.00,$$
$$\{a \to \ <name\ of\ area>, \quad t \to \ <time>, \quad b \to AREA <name\ of\ area> IS <state>\}>.$$

As seen,

$$s[\delta] = AREA <name\ of\ area> IS\ SMOKED\ AT <time>,$$

$$s'[\delta] = AREA <name\ of\ area> IS <state> AT\ 15.00,$$

$$y = \inf\{s[\delta], s'[\delta]\} = AREA <name\ of\ area> IS\ SMOKED\ AT\ 15.00,$$

and thus

$$\overline{\delta} = \{a \to \ <name\ of\ area>, \qquad t \to 15.00,$$

$$b \to AREA <name\ of\ area> IS\ SMOKED\}.\blacksquare$$

Now we may return to DBI and introduce the so-called term data manipulation language (*TDML*), being the set of the so-called augmented terms $<s, d>$, where s is the term and d is the SF-substitution. M-semantics of this language is similar to Eqs. (49), (54), and (55) and is obtained by replacement of *SF y* by couple $<s, d>$:

$$A_{t+1}^{s,d} = \left\{x | x \in X_t \& s[d] \overset{*}{\Rightarrow} x\right\}, \tag{88}$$

$$\overline{A}_{t+1}^{s,d} = \{x | x \in X_t \& \exists \inf\{s[d], x\}\}, \tag{89}$$

$$\overline{\overline{A}}_{t+1}^{s,d} = \{\inf\{s[d], x\} | x \in X_t \& \exists \inf\{s[d], x\}\}. \tag{90}$$

Moreover, from now we may use augmented terms or even their sets as N-facts. Corresponding equations, which describe M-semantics of TDML, much more useful from the practical point of view, are as follows:

$$A_{t+1}^{s,d} = \left\{ <\bar{s},\bar{d}> | <\bar{s},\bar{d}> \in X_t \& s[d] \overset{*}{\Rightarrow} \bar{s}\left[\bar{d}\right] \right\}, \tag{91}$$

$$\overline{A}_{t+1}^{s,d} = \left\{ <\bar{s},\bar{d}> | <\bar{s},\bar{d}> \in X_t \& \exists \inf\left\{s[d], \bar{s}\left[\bar{d}\right]\right\} \right\}, \tag{92}$$

$$\overline{\overline{A}}_{t+1}^{s,d} = \left\{ D\left[s = \bar{s}, d \cup \bar{d}\right] | <\bar{s},\bar{d}> \in X_t \right\}. \tag{93}$$

As may be seen, the last definition provides the most informative reply, containing maximal unifiers of WECFL, each corresponding N-fact, entering BDI.

The concerned reader may find the detailed consideration of WECFL, DBI algorithmics (including N-facts fusion), and key theoretical issues of SDB/DBI internal organization, providing associative access to the stored data as well as their compression, in [38–40].

All the said about TDML is sufficient for consideration of already mentioned knowledge representation, called augmented Post systems, being core of the deductive capabilities of "Set of Strings" Framework. APS are described in the separate chapter of this book.

Author details

Igor Sheremet
Financial University under the Government of Russian Federation, Moscow, Russia

*Address all correspondence to: sheremet@rfbr.ru

References

[1] Smolan R, Erwitt J. The Human Face of Big Data. Abingdon, UK: Marston Book Services; 2012. p. 224

[2] Roberts FS. What is Big Data and how has it changed. In: Invited Talk at International Conference on Data Intensive Systems Analysis for Geohazard Studies; July 18–21, 2016; Sochi, Russia; 2016

[3] Chen M, Mao S, Zhang Y, Leung VCM. Big Data: Related Technologies, Challenges, and Future Prospects. NY: Springer; 2014. p. 100. DOI: 10.1007/978-3-319-06245-7

[4] Mayer-Schonberger V, Cukier R. Big Data: A Revolution that Will Transform how we Live, Work, and Think. Canada: Eamon Dolan/Houghton Mifflin Harcourt; 2013. p. 242

[5] Labrinidis A, Jagadish HV. Challenges and opportunities with big data. Proceedings of the VLDB Endowment. 2012;**5**(12):2032-2033

[6] Maheshwari A. Data Analytics Made Accessible. Seattle, WA: Amazon Digital Services; 2015. p. 156

[7] Marz N, Warren J. Big Data: Principles and Best Practices of Scalable Real-time Data Systems. NY: Manning Publications; 2015. p. 328

[8] Codd EF. The Relational Model for Database Management: Version 2. Boston, MA: Addison Wesley; 1990. p. 567

[9] Date CJ. An Introduction to Database Systems. Pearson; 2003. p. 1034

[10] Sumathi S, Esakkirajan S. Fundamentals of Relational Database Management Systems. NY: Springer; 2007. p. 776

[11] Mensah K. Oracle Database Programming Using Java and Web Services. Amsterdam: Elsevier; 2006. p. 1120

[12] Vaswani V. MySQL Database Usage & Administration. NY: McGraw Hill; 2009. p. 368

[13] Pendyala V. Veracity of Big Data: Machine Learning and Other Approaches to Verifying Truthfulness. NY: Apress; 2018. p. 177

[14] McNeill C. Veracity: The Most Important "V" of Big Data. GutCheck. January 23, 2018. Available from: www.gutcheckit.com/blog/veracity-big-data

[15] Normandeau K. Beyond Volume, Variety and Velocity is the Issue of Big Data Veracity. Inside Big Data. September 12, 2003. Available from: www.insidebigdata.com/2013/09/12/beyond-volume-variety-velocity-issue-big-data-veracity

[16] Walker M. Data Veracity. Data Science Central. November 28, 2012. Available from: www.datasciencecentral.com/profiles/blogs/data-veracity

[17] Siewert SB. Big Data in the Cloud. Data Velocity, Volume, Variety, Veracity. Available from: www.ibm.com/develoerworks/library/bd-bigdatacloud

[18] Tan P-N, Steinbach M, Kumar V. Introduction to Data Mining. London, UK: Pearson/Addison Wesley; 2005. p. 400

[19] Witten IH, Frank E, Hall MA. Data Mining: Practical Machine Learning Tools and Techniques. 3rd ed. Burlington, MA: Morgan Kaufmann; 2011. p. 664

[20] Data Mining. Project Report Online. December 27, 2017. Available from: www.projectreportonline.com/data-mining/

[21] Adriaans P, Zantinge D. Data Mining. London, UK: Pearson; 2005. p. 458

[22] Chomsky N. Syntactic Structures. 2nd ed. Berlin, New York: Mouton de Gruyter; 2002. p. 117

[23] Post EL. Formal reductions of the general combinatorial decision problem. American Journal of Mathematics. 1943; **65**:197-215

[24] Teeling M. What is Data Integrity? Learn How to Ensure Database Data Integrity via Checks, Tests, & Best Practices. May 14, 2012. Available from: www.veracode.com/blog/2012/05/what-is-data-integrity

[25] Haloin T, Morgan T. Information Modelling and Relational Databases. Burlington, MA: Morgan Kaufmann; 2010. p. 976

[26] Date C. Database Design and Relation Theory: Normal Forms and all that Jazz. Sebastopol, CA: O'Reilly Media, Inc.; 2012. p. 260

[27] Silberschatz A, Korth H, Sudarshan S. Database Systems Concepts. NY: McGraw Hill; 2012

[28] Garcia-Molina H, Ullman JD, Widom J. Database Systems: The Complete Book. 2nd ed. London, UK: Pearson Prentice Hall; 2009

[29] Fagin R, Vardi MY. The theory of data dependencies—A survey. In: Anshel M, Gewirtz W, editors. Mathematics of Information Processing. Proceedings of Symposia in Applied Mathematics. Vol. 34. 1986. pp. 19-71

[30] Demetrovics J, Libkin L, Muchnik IB. Functional dependencies in relational databases: A lattice point of view. Discrete Applied Mathematics. 1992;**40**(2):155-185. DOI: 10.1016/0166-218X(92)90028-9

[31] Megid YA, El-Tazi N, Fahmy A. Using functional dependencies in conversion of relational databases to graph databases. In: DEXA 2018: Database and Expert Systems Applications. Lecture Notes in Computer Science. Vol. 11030; 2018. pp. 350-357

[32] Wiil UK. Experience with hyperbase: A hypertext database supporting collaborative work. SIGMOD Record. 1993;**22**(4):19-25

[33] De Bra P, Pechenitzkiy M. Dynamic and adaptive hypertext: Generic frameworks, approaches and techniques. In: Proceedings of the 20th ACM Conference on Hypertext and Hypermedia. 2009. pp. 387-388. DOI: 10.1145/1557914.1558003

[34] Bagchi A, Lahoti G. Relating web pages to enable information-gathering tasks. In: Proceedings of the 20th ACM Conference on Hypertext and Hypermedia. 2009. pp. 109-118. DOI: 10.1145/1557914.1557935

[35] Lescovec J. Human navigation in networks. In: Proceedings of the 23th ACM Conference on Hypertext and Hypermedia. 2012. pp. 143-144. DOI: 10.1145/2309996.2310020

[36] Hall W. From hypertext to linked data: The ever evolving web. In: Proceedings of the 22th ACM Conference on Hypertext and Hypermedia. 2011. pp. 3-4. DOI: 10.1145/1995996.1995969

[37] Hara Y, Botafogo R. Hypermedia databases: a specification and formal language. In: Proceedings of Database and Expert Systems Applications DEXA'94; Athens, Greece; September 7–9, 1994. pp. 521-530. DOI: 10.1007/3-540-58435-8-218

[38] Sheremet IA. Intelligent Software Environments for Information Processing Systems. Moscow: Nauka; 1994. p. 544. (In Russian)

[39] Sheremet IA. Grammatical Codings. Hannover: EANS; 2012. p. 54

[40] Sheremet IA. Augmented Post Systems: The Mathematical Framework for Data and Knowledge Engineering in Network-Centric Environment. Berlin: EANS; 2013. p. 395

[41] Sheremet I. Data and knowledge bases with incomplete information in a "Set of Strings" framework. International Journal of Engineering and Applied Sciences. 2016;**3**(3):90-103

[42] McCreary D, Kelly A. Making Sense of NoSQL. A Guide for Managers and the Rest of us. NY: Manning Publications; 2013. p. 312

[43] Tiwari S. Professional NoSQL. Birmingham, UK: Packt Publishing; 2011. p. 384

[44] Moniruzzaman AB, Hossain SA. NoSQL Database: New Era of Databases for Big Data Analytics—Classifications, Characteristics and Comparison; 2013. arXiv: 1307.0191

[45] Kessin Z. Building Web Applications with Erlang. Sebastopol, CA: O'Reilly Media, Inc.; 2012. p. 156

[46] Oracle Berkeley DB. Available from: www.oracle.com/technetwork/database/database-technologies/berkeleydb/overview/index.html

[47] Redis. Available from: www.redis.io

[48] Cunningham L. World's Largest Database Runs on Postgres? Available from: https://it.toolbox.com/blogs/lewiscunningham/worlds-largest-database-runs-on-postgres-052808

[49] Meduna A. Formal Languages and Computation: Models and their Applications. London, UK: CRC Press; 2014. p. 315

[50] Kolmogorov AN. Three approaches to the definition of notion "quantity of information". In: The Selectas. Vol. 3 "Information Theory and Theory of Algorithms". Moscow: Nauka; 2005. (In Russian)

[51] Sheremet IA. Word Equations on Context-Free Languages. Hannover: EANS; 2011. p. 44

[52] Schultz KU, editor. Word equations and related topics. Lecture Notes in Computer Science. Vol. 572. NY: Springer Verlag; 1992. p. 256

[53] Lothaire M. Algebraic Combinatorics on Words. Cambridge, UK: Cambridge University Press; 2002. p. 504

[54] Makanin GS. The problem of solvability of equations in a free semigroup. Mathematics of the USSR-Sbornik. 1977;**32**(2):129-198

[55] Makanin GS. Equations in a free group. Mathematics of the USSR-Izvestiya. 1983;**21**(3):483-546

[56] Makanin GS. Solvability of universal and positive theories in a free group. Mathematics of the USSR-Izvestiya. 1985;**25**(1):75-83

Chapter 7

Investigation of Fuzzy Inductive Modeling Method in Forecasting Problems

Yu. Zaychenko and Helen Zaychenko

Abstract

This paper is devoted to the investigation and application of fuzzy inductive modeling method group method of data handling (GMDH) in problems of forecasting in the financial sphere. GMDH method belongs to self-organizing methods and allows to discover internal hidden laws in the appropriate object area. The advantage of GMDH algorithms is the possibility of constructing optimal models. In the generalization of GMDH in case of uncertainty, new method fuzzy GMDH is described which enables to construct fuzzy models almost automatically. The algorithms of fuzzy GMDH for different membership functions are considered. The extensions of fuzzy GMDH for different partial descriptions—orthogonal polynomials of Chebyshev and trigonometric polynomials of Fourier—are considered. The problem of adaptation of fuzzy models obtained by FGMDH is considered, and the corresponding adaptation algorithm is described. The experimental investigations of the suggested FGMDH in the problem of forecasting macroeconomic indicators of Ukraine are carried out, and comparison with classic GMDH and neural network BP is performed.

Keywords: fuzzy GMDH, orthogonal partial descriptions, model adaptation, forecasting

1. Introduction

One of the most important problems in the sphere of economy and finance is the problem of forecasting economic and financial processes. The distinguishing properties of these processes are the following:

1. The form of functional dependence is unknown, and only model class is determined.
2. Short data samples.
3. Time series $x_i(t)$ in general case is nonstationary.

In this case the application of traditional methods of statistical analysis (e.g., regression analysis) is impossible, and it's necessary to apply methods based on computational intelligence (CI). To this class belongs group method of data handling (GMDH) developed by Ivakhnenko [1, 2] and extended by his colleagues. GMDH method belongs to self-organizing methods and allows to discover hidden laws in the appropriate object area. The advantage of GMDH algorithms is the capability of constructing optimal models.

But classic GMDH has the following shortcomings:

1. GMDH utilizes least squared method (LSM) for finding the model coefficients, but matrix of linear equations may be close to degenerate, and the corresponding solution may appear non-stable. Therefore, the special methods for its regularization should be used.

2. GMDH doesn't work in the case of qualitative or fuzzy input data.

Therefore in the last 10 years, the new variant of GMDH—fuzzy GMDH—was developed and extended which may work with fuzzy input data and is free of classical GMDH drawbacks [3–5].

Fuzzy GMDH is also based on the same principles as classical GMDH but construct fuzzy models.

The main goals of this paper are to investigate different modifications of FGMDH, analyze their properties, and investigate its efficiency as compared with classical GMDH in forecasting problems.

2. Problem formulation

A set of initial data is given inclusive of input variables $\{X(1), X(2), \ldots, X(N)\}$ and output variables $\{Y(1), Y(2), \ldots, Y(N)\}$, where $X = [x_1, x_2, \ldots, x_n]$ is -n-tuple vector, N is a number of observations, and input data may be incomplete or fuzzy, in particular given in interval form. The task is to construct an adequate fuzzy forecasting model $Y = F(x_1, x_2, \ldots, x_n)$, and besides, the obtained model should have the minimal complexity.

2.1 Principal ideas of GMDH: fuzzy model construction

As it well known, the drawbacks of GMDH are the following [3, 4]:

- GMDH utilizes LSM for finding the model coefficients, but matrix of linear equations may be close to degenerate, and the corresponding solution may appear non-stable and very volatile. Therefore, the special regularization methods should be applied.

- After application of GMDH point-wise estimations is obtained, but in many cases, it's desirable to find interval value for coefficient estimates.

- GMDH doesn't work in the case of incomplete, qualitative, or fuzzy input data.

Therefore, in the last 10 years, the new variant of GMDH—fuzzy GMDH—was developed and improved which may work with fuzzy and qualitative input data and is free of classical GMDH drawbacks [3].

As it is well known, GMDH method is based on the following principles [1–3]:

1. The principle of multiplicity of models

2. The principle of external complement which means that the whole sample should be divided into two parts—training subsample and test subsample

3. The principle of self-organization

4. The principle of freedom of choice

Fuzzy GMDH is also based on these principles but construct fuzzy models. Let's consider its main ideas.

In works [3–5], the linear interval regression model was considered:

$$Y = A_0 Z_0 + A_1 Z_1 + \dots + A_n Z_n \tag{1}$$

where A_i is a fuzzy number of triangular form described by a pair of parameters $A_i = (\alpha_i, c_i)$, where α_i is interval center, c_i is its width, and $c_i \geq 0$, Zi is the input variables.

Then Y is a fuzzy number, parameters of which are determined as follows:

The interval center

$$\alpha_y = \sum \alpha_i z_i = \alpha^T \cdot z \tag{2}$$

The interval width

$$c_y = \sum c_i \cdot |z_i| = c^T |z| \tag{3}$$

For example, for the partial description (PD) of the kind

$$f(x_i, x_j) = A_0 + A_1 x_i + A_2 x_j + A_3 x_i x_j + A_4 x_i^2 + A_5 x_j^2 \tag{4}$$

it's necessary to substitute in the general model (1)

$z_0 = 1\, z_1 = x_i\, z_2 = x_j\, z_3 = x_i x_j\, z_4 = x_i^2\, z_5 = x_j^2$.

Let the training sample be $\{z_1, z_2, \dots, z_M\}$, $\{y_1, y_2, \dots, y_M\}$. Then for the model (1) to be adequate, it's necessary to find such parameters $(\alpha_i, c_i)\ i = \overline{1, n}$, which satisfy the following inequalities:

$$\begin{cases} \alpha^T z_k - c^T \cdot |z_k| \leq y_k \\ \alpha^T z_k + c^T \cdot |z_k| \geq y_k \end{cases}, \quad k = \overline{1, M} \tag{5}$$

Let's formulate the basic requirements for the linear interval model of a kind (4).

It's necessary to find such values of the parameters (α_i, c_i) of fuzzy coefficients for which:

1. Real values of the observed outputs y_k should drop in the estimated interval for Y_k.

2. The total width of the estimated interval for all sample points should be minimal.

These requirements lead to the following linear programming (LP) problem [3, 4]:

$$\min(C_0 \cdot M + C_1 \sum_{k=1}^{M} |x_{ki}| + C_2 \sum_{k=1}^{M} |x_{kj}| + C_3 \sum_{k=1}^{M} |x_{ki}x_{kj}| + \\ + C_4 \sum_{k=1}^{M} |x_{ki}^2| + C_5 \sum_{k=1}^{M} |x_{kj}^2| \quad (6)$$

under constraints

$$a_0 + a_1 x_{ki} + a_2 x_{kj} + a_3 x_{ki} x_{kj} + a_4 x_{ki}^2 + a_5 x_{kj}^2 - (C_0 + C_1|x_{ki}| + C_2|x_{kj}| + \\ + C_3|x_{ki}x_{kj}| + C_4|x_{ki}^2| + C_5|x_{kj}^2|) \le y_k \quad (7)$$

$$a_0 + a_1 x_{ki} + a_2 x_{kj} + a_3 x_{ki} x_{kj} + a_4 x_{ki}^2 + a_5 x_{kj}^2 + (C_0 + C_1|x_{ki}| + C_2|x_{kj}| + \\ + C_3|x_{ki}x_{kj}| + C_4|x_{ki}^2| + C_5|x_{kj}^2|) \ge y_k, \quad (8)$$

$$C_p \ge 0, \;\; p = 0, 5 \; k = \overline{1, M}$$

where k is a number of a point.

As one can easily see, the task (6)–(8) is a LP problem. However, the inconvenience of the model (6)–(8) for the application of standard LP methods is that there are no constraints of non-negativity for variables α_i. Therefore for its solution, it's reasonable to pass to the dual LP problem by introducing dual variables $\{\delta_k\}$ and $\{\delta_{k+M}\}$, $k = \overline{1, M}$. Using simplex method after finding the optimal solution for the dual problem, the optimal solutions (α_i, c_i) of the initial direct problem will be also found.

3. Description of fuzzy GMDH algorithm

Let's present the brief description of the algorithm FGMDH [3, 4].

1. Choose the general model type by which the sought dependence will be described.

2. Choose the external criterion of optimality (criterion of regularity or non-biasedness).

3. Choose the type of partial descriptions (e.g., linear or quadratic one).

4. Divide the sample into training N_{train} and test N_{test} subsamples.

5. Put zero values to the counter of model number k and to the counter of iteration number r.

6. Generate a new partial model f_k (4) using the training sample. Solve the LP problem (6)–(8), and find the values of parameters α_i, c_i.

7. Calculate the value of external criterion ($N_{ubk}^{(r)}$ or $\delta_k^{(2)}(r)$) at the test sample.

8. $k = k + 1$. If $k > C_N^2$ for $r = 1$ or $k > C_F^2$ for $r > 1$, then $k = 1$, $r = r + 1$, and go to step 9; otherwise go to step 6.

9. Calculate the best value of the criterion for models of rth iteration $\delta^{(2)}(r)$ or $N_{ub}^{(r)}$. If $r = 1$, then go to step 6; otherwise go to step 10.

10. If $|N_{ub}(r) - N_{ub}(r-1)| \leq \varepsilon$ or $\delta^{(2)}(r) \geq \delta^{(2)}(r-1)$, then go to step 11; otherwise select F best models, assign $r = r + 1$ and $k = 1$, go to step 6, and execute $(r + 1)$th iteration.

11. Select the best model of the previous iteration using external criterion and moving back by its connections and successively passing all the previous rows find analytical form the constructed model.

4. Analysis of different membership functions

In the first papers devoted to fuzzy GMDH [3], the triangular membership functions (MFs) were considered. But as fuzzy numbers may also have the other kinds of MF, it's important to consider the other classes of MF in the problems of modeling using FGMDH. In paper [4] fuzzy models with Gaussian and bell-shaped MFs were investigated.

Consider a fuzzy set with Gaussian MF:

$$\mu_B(x) = e^{-\frac{1}{2}\frac{(x-1)^2}{c^2}} \tag{9}$$

Let the linear interval model for partial description of FGMDH take the form (4). Then the problem is formulated as follows:

Find such fuzzy numbers B_i, with parameters (a_i, c_i), that:

- The observation y_k would belong to a given estimate interval for the set $Y(k)$ with degree not less than α, $0 < \alpha < 1$.

- The width of estimated interval of the degree α would be minimal.

In [4, 6] it was shown that the problem of finding optimal fuzzy model will be finally transformed to the following LP problem:

$$\begin{aligned} \min(C_0 \cdot M + C_1 \sum_{k=1}^{M} |x_{ki}| + C_2 \sum_{k=1}^{M} |x_{kj}| + C_3 \sum_{k=1}^{M} |x_{ki}x_{kj}| + \\ + C_4 \sum_{k=1}^{M} |x_{ki}^2| + C_5 \sum_{k=1}^{M} |x_{kj}^2| \end{aligned} \tag{10}$$

under constraints

$$\left.\begin{aligned} a_0 + a_1 x_{ki} + \ldots + a_5 x_{kj}^2 + \left(C_0 + C_1|x_{ki}| + \ldots + C_5\left|x_{kj}^2\right|\right) \cdot \sqrt{-2\ln\alpha} \geq y_k \\ a_0 + a_1 x_{ki} + \ldots + a_5 x_{kj}^2 - \left(C_0 + C_1|x_{ki}| + \ldots + C_5\left|x_{kj}^2\right|\right) \cdot \sqrt{-2\ln\alpha} \leq y_k \end{aligned}\right\} k = \overline{1, M} \tag{11}$$

To solve this problem like the case with triangular MF, it's reasonable to pass to the dual LP problem of the form

$$\max\left(\sum_{k=1}^{M} y_k \cdot \delta_{k+\mathrm{M}} - \sum_{k=1}^{M} y_k \cdot \delta_k\right) \tag{12}$$

with constraints of equalities and inequalities

$$\sum_{k=1}^{M}\delta_{k+M} - \sum_{k=1}^{M}\delta_k = 0,$$
$$\sum_{k=1}^{M}X_{ki}\cdot\delta_{k+M} - \sum_{k=1}^{M}X_{ki_i}\cdot\delta_k = 0 \tag{13}$$
$$\cdots$$
$$\sum_{k=1}^{M}X_{kj}{}^2\cdot\delta_{k+M} - \sum_{k=1}^{M}X_{k_j}^2\cdot\delta_k = 0$$

$$\left.\begin{array}{c}\sum_{k=1}^{M}\delta_k + \sum_{k=1}^{M}\delta_{k+M} \le \frac{M}{\sqrt{-2\ln\alpha}}\\ \sum_{k=1}^{M}|X_{ki}|\cdot\delta_{k+M} + \sum_{k=1}^{M}|X_{ki}|\cdot\delta_k \le \frac{\sum_{k=1}^{M}|X_{ki}|}{\sqrt{-2\ln\alpha}}\\ \cdots\\ \sum_{k=1}^{M}\left|X_{kj}^2\right|\cdot\delta_{k+M} + \sum_{k=1}^{M}\left|X_{kj}^2\right|\cdot\delta_k \le \frac{\sum_{k=1}^{M}\left|X_{kj}^2\right|}{\sqrt{-2\ln\alpha}}\end{array}\right\} \tag{14}$$

$$\delta_k \ge 0,\ k = \overline{1, 2M} \tag{15}$$

Analyzing dual LP program (12)–(15), it's easy to notice that this problem is always solvable as there trivial solution $\delta_k = 1\ k = \overline{1, 2M}$ always exists. Therefore the initial problem (10) and (11) also always has solutions with any data.

Thus, fuzzy GMDH allows to construct fuzzy models and has the following advantages:

1. The problem of optimal model determination is transformed to the problem of linear programming, which is always solvable.

2. As the result of method work, the interval regression model is being built.

5. Fuzzy GMDH with different partial descriptions: orthogonal polynomials

As it is well known from the general GMDH theory, model pretenders are generated on the base of so-called partial description—elementary models of two variables. Usually as partial descriptions, linear or quadratic polynomials are used. The alternative to this class of models is application of orthogonal polynomials. The choice of orthogonal polynomials as partial descriptions is determined by the following advantages:

- Due to orthogonal property, the determination of polynomial coefficients goes faster than for non-orthogonal polynomials.

- The coefficients of polynomial approximating equation don't depend on the real degree of initial polynomial model, so if a priori the real polynomial degree is not known, one may calculate the polynomials of various degrees, and by this property the coefficients obtained for polynomials of lower degrees remain the same after transfer to higher polynomial degrees. This property is the most important during investigation of real degree of approximating polynomial.

5.1 Chebyshev's orthogonal polynomials

Chebyshev's orthogonal polynomials in general case have the following form [5]:

$$F_\nu(\xi) = T_\nu(\xi) = \cos(\nu \cdot \arccos\xi), \quad -1 \le \xi \le 1 \tag{16}$$

These polynomials have the following orthogonality property:

$$\int_{-1}^{1} \frac{T_\mu(\xi) T_\nu(\xi) d\xi}{\sqrt{1-\xi^2}} = \begin{cases} 0 \text{ if } \mu \ne \xi; \\ \frac{\pi}{2} \text{ if } \mu = \xi \ne 0; \\ \pi \text{ if } \mu = \xi = 0. \end{cases} \tag{17}$$

where $\sqrt{1-\xi^2}$ is a weighting coefficient $\omega(\xi)$ in the Eq. (17).

The approximating Chebyshev's orthogonal polynomial for $\bar{y}$ is obtained on the base of function S minimization:

$$S = \int_{-1}^{1} \omega(\xi) \left(y(\xi) - \sum_{i=0}^{m} b_i T_i(\xi) \right)^2 d\xi \tag{18}$$

where from (18) we obtain the following expression for coefficients:

$$b_k = \begin{cases} \frac{1}{\pi} \int_{-1}^{1} \frac{y(\xi)}{\sqrt{1-\xi^2}} d\xi, & k = 0 \\ \frac{2}{\pi} \int_{-1}^{1} \frac{y(\xi) T_k(\xi)}{\sqrt{1-\xi^2}} d\xi, & k \ne 0 \end{cases} \tag{19}$$

Hence, the approximating equation takes the form

$$\bar{y}(\xi) = \sum_{k=0}^{m} b_k T_k(\xi) \tag{20}$$

As it may be readily seen from the presented expressions, coefficient b_k in Eq. (19) doesn't depend on the choice of degree m. Thus, the variable m doesn't demand recalculation of b_j, $\forall j \le m$, while such recalculation is necessary for non-orthogonal approximation.

The best degree m^* of approximating may be obtained on the base of hypothesis that investigation results $y(i)$, $i = 1, 2, \ldots, r$ have independent Gaussian distribution in the bounds of some polynomial function $\bar{y}$ of definite degree, for example, $m^* + \mu$, where

$$\bar{y}_{m^*+\mu}(x_i) = \sum_{j=0}^{m^*+\mu} b_j x_i^j \tag{21}$$

and a dispersion σ^2 of distribution y-$\bar{y}$ don't depend on μ.

It's clear that for very small m (m =0,1,2,...), σ_m^2 decreases as m grows.

As in accordance with previously formulated hypothesis, dispersion doesn't depend on μ; therefore, the best degree m^* is a minimal m, for which $\sigma_m \cong \sigma_{m+1}$.

For determining m^* it's necessary to calculate the approximating polynomials of various degrees. As coefficients b_j in Eq. (20) don't depend on μ, the determination of the best degree of polynomial is accelerated.

Let us have the forecasted variable Y and input variables $x_1, x_2, \ldots x_n$. Let's search the relation between them in the following form:

$$Y = A_1 f_1(x_1) + A_2 f_2(x_2) + \ldots + A_n f_n(x_n) \tag{22}$$

where A_i is a fuzzy number of triangular type given as $A_i = (\alpha_i, c_i)$, functions f_i are determined so [5, 6]

$$f_i(x_i) = \sum_{j=0}^{m_i} b_{ij} T_j(x_i) \tag{23}$$

The degree m_i of function f_i is determined using hypothesis defined in the preceding section. So if we denote $z_i = f_i(x_i)$, we'll get linear interval model in its classical form.

5.2 Investigation of trigonometric polynomials as partial descriptions

Let function $f(x)$ be periodic with period 2π defined at the interval $[-\pi, \pi]$, and its derivative $f'(x)$ is also defined at the interval $[-\pi, \pi]$. Then the following equality holds

$$S(x) = f(x); \forall x \in [-\pi, \pi] \tag{24}$$

where

$$S(x) = \frac{a_0}{2} + \sum_{j=1} (a_j \cos(jx) + b_j \sin(jx)) \tag{25}$$

Coefficients a_j and b_j are calculated by Euler formulas:

$$\begin{aligned} a_j &= \frac{1}{\pi} \int_{-\pi}^{\pi} f(x) \cos(jx) dx; \\ b_j &= \frac{1}{\pi} \int_{-\pi}^{\pi} f(x) \sin(jx) dx; \end{aligned} \tag{26}$$

5.3 Definition

A trigonometric polynomial of the degree M is called the following polynomial:

$$T_M(x) = \frac{a_0}{2} + \sum_{j=1}^{M} (a_j \cos(jx) + b_j \sin(jx)) \tag{27}$$

The following theorem is true stating that exists such M, where $2M < N$, which minimizes the following criterion:

$$\sum_{j=1}^{N} (f(x_i) - T_M(x_i))^2 \tag{28}$$

Hence the coefficients of corresponding trigonometric polynomial are determined by formulas

$$a_j = \frac{2}{N}\sum_{i=1}^{N} f(x_i)\cos\left(jx_i\right);$$
$$b_j = \frac{2}{N}\sum_{i=1}^{N} f(x_i)\sin\left(jx_i\right); \tag{29}$$

Let it be the forecasted variable Y and input variables $x_1, x_2, \ldots x_n$. Let's search the dependence among them in the form

$$Y = A_1 f_1(x_1) + A_2 f_2(x_2) + \ldots + A_n f_n(x_n) \tag{30}$$

where A_i is a fuzzy number of triangular type given as $A_i = (\alpha_i, c_i)$, functions f_i are determined in such a way:

$$f_i(x_i) = T_{M_i}(x_i) \tag{31}$$

The degree M_i of function f_i is determined by the theorem described in the preceding section. Therefore, if we assign $z_i = f_i(x_i)$, the linear interval model will be obtained in its classical form.

6. Adaptation of fuzzy GMDH models

While forecasting by self-organizing methods (fuzzy GMDH, in particular), the problem of adaptation arises in the case of the training sample size increase when it's needed to correct the obtained model in accordance with new available data. Taking into account new information obtained while forecasting adaptation may be done by two approaches. The *first* one is to correct parameters of a forecasting model with new data assuming that model structure didn't change. The second approach consists in adaptation of not only model parameters but its optimal structure as well. This way demands the repetitive use of full GMDH algorithm and is connected with huge volume of calculations.

The second approach is used if adaptation of parameters doesn't provide good forecast and the new real output values don't drop in the calculated interval for its estimate.

In our work the first approach is used based on adaptation of FGMDH model parameters with new available data. Here the recursive identification methods are preferably used, especially the recursive LSM. In this method the parameter estimates at the next step are determined on the base of estimates at the previous step, model error, and some information matrix which is modified during all estimation process and therefore contains data which may be used at the next steps of adaptation process [5].

Hence, model coefficient adaptation will be simplified substantially. If we store information matrix obtained while identification of optimal model using fuzzy GMDH, then for model parameters adaptation, it will be enough to fulfill only one iteration by recursive LSM method.

6.1 The application of recurrent LSM for model coefficients adaptation

Consider the following model:

$$y(k) = \theta^T \Psi(k) + v(k) \tag{32}$$

where $y(k)$ is a dependent (output) variable, $\Psi(k)$ is a measurement vector, $v(k)$ are random disturbances, and θ is a parameter vector to be estimated.

The parameters estimate θ at the step N is performed due to such formula [5, 6]:

$$\widehat{\theta}\ (N) = \widehat{\theta}\ (N-1) + \gamma(N)\left[y(N) - \widehat{\theta}\ T(N-1)\Psi(N)\right] \tag{33}$$

where $\gamma(N)$ is a coefficient vector which is determined by formula

$$\gamma(N) = \frac{P(N-1)\Psi(N)}{1 + \Psi^T(N)P(N-1)\Psi(N)} \tag{34}$$

where $P(N-1)$ is so-called an information matrix, determined by formula

$$P(N-1) = P(N-2) - \frac{P(N-2)\Psi(N-1)\Psi^T(N-1)P(N-2)}{1 + \Psi^T(N-1)P(N-2)\Psi(N-1)} \tag{35}$$

As one can easily see in (35), the information matrix may be obtained independent on parameter estimation process and parallel to it. The adaptation of two parameter vectors $\theta_1^T = [\alpha_1, \ldots, \alpha_m]$; $\theta_2^T = [C_1, \ldots, C_m]$; is performed in such a way using the formulas (35)

$$\begin{aligned} \widehat{\theta}_1(N) &= \widehat{\theta}_1(N-1) + \gamma_1(N)\left[y(N) - \widehat{\theta}_1^T(N-1)\Psi_1(N)\right] \\ \widehat{\theta}_2(N) &= \widehat{\theta}_2(N-1) + \gamma_2(N)\left[y_c(N) - \widehat{\theta}_2^T(N-1)\Psi_2(N)\right] \\ y_c(N) &= |y(N) - \theta_1^T(N-1)\Psi_1(N)| \end{aligned} \tag{36}$$

where $\Psi_1^T = [z_1, \ldots, z_m]$; $\Psi_2^T = [|z_1|, \ldots, |z_m|]$.

7. Experimental investigations of FGMDH in forecasting

The goal of experiments was the forecasting of macroeconomic indicators of Ukraine and estimating of efficiency of suggested FGMDH. In experiments, the database was utilized which contains monthly values of 24 macroeconomic indicators of Ukrainian economy, since July 1995 till 2013. As forecasting variables consumer price index (CPI) and gross national product (GNP) were chosen.

While constructing forecasting models, the technique of sliding window was utilized, whose size was determined automatically by regression analysis. For determination of input variables significant for forecasting, the methods of regression analysis were also used.

The following experiments were performed:

1. Forecasting model construction with application of different membership functions: triangular, Gaussian, and bell-wise one

2. For macroeconomic indicators (CPI and GNP) forecasting the construction of forecasting models using different partial descriptions—classic Chebyshev's polynomials and trigonometric polynomials.

3. For adaptation of models, the algorithm of stochastic approximation and recurrent least squared method (RLSM) were applied.

4. Comparative analysis of the suggested algorithms with classic GMDH and neural networks (NN), in particular neural network backpropagation, was performed.

7.1 Comparison of different membership functions

The experimental investigations of fuzzy forecasting models were carried out with following MF: triangular, Gaussian, and bell-wise. As accuracy criteria RMSE was chosen. RMSE values while forecasting CPI are presented in **Figure 1**.

As one can see, the most efficient for constructing linear interval models is application of bell-wise membership functions for fuzzy coefficients, on the second place are Gaussian MFs, and the worst forecasting accuracy was achieved with triangular MF. In the next experiment, the task was to forecast GPD values.

In **Figure 2** the obtained RMSE values for forecasting GNP are presented.

As one can see, the results are practically the same as in the previous experiment. The best accuracy was attained with bell-wise MF.

7.2 Comparison of different partial descriptions

In the next series of experiments, the investigations of FGMDH models with the following partial descriptions were carried out: quadratic polynomials, Chebyshev's polynomials, trigonometric polynomials, and ARIMA models. In **Figure 3** accuracy of forecasting PCI is presented with different PD.

As we can see, the best results are obtained with models which use trigonometric polynomials as PD. Somewhat worse are results with classic quadratic polynomials. And the worst turned out to be ARIMA models as PD. It may be explained by the fact that ARIMA models are functions of one variable. That is a serious drawback of such models.

7.3 Comparison of crisp and fuzzy GMDH

For more comprehensive efficiency comparison of crisp and fuzzy GMDH, existing implementation of GMDH was extended by inclusion of new types of PD orthogonal polynomials: Chebyshev's and trigonometric and ARIMA models as PD.

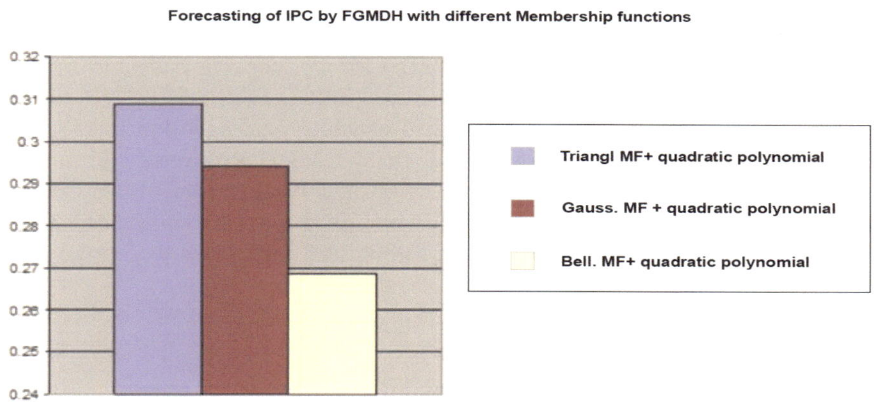

Figure 1.
Forecasting accuracy of PCI with different MF.

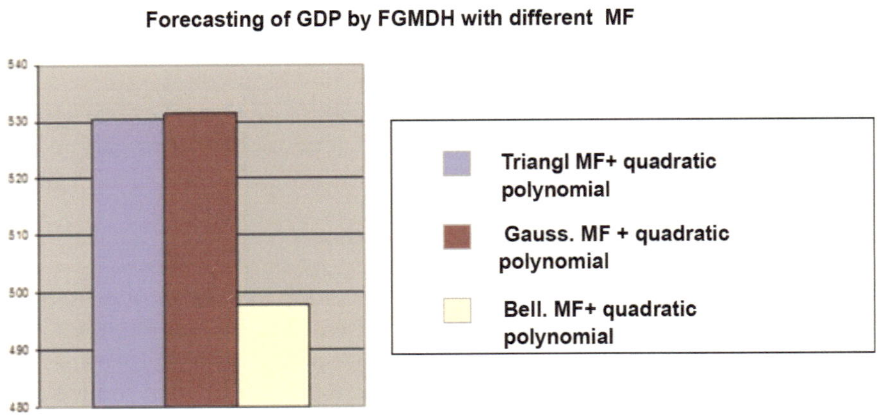

Figure 2.
Forecasting accuracy of GNP with different MF.

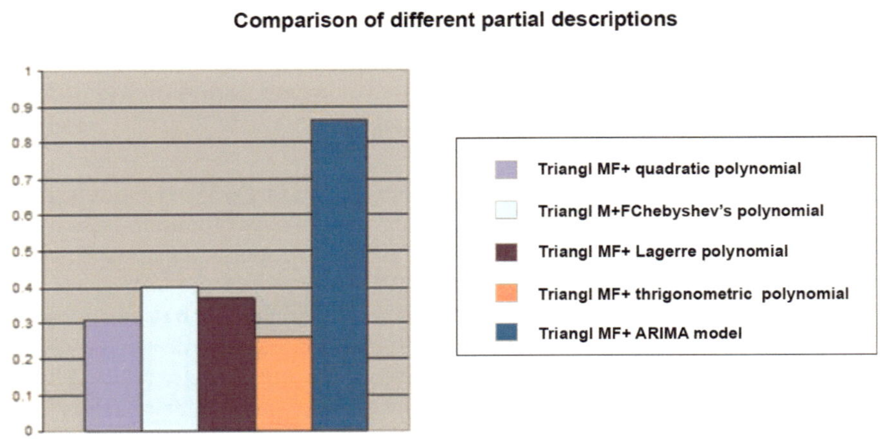

Figure 3.
Forecasting accuracy of PCI for different PD.

As adaptation algorithm stochastic approximation and recurrent LSM were implemented. In **Figures 4** and **5**, the mean RMSE values for crisp and fuzzy GMDH in the whole range of data variation are presented for different types of PD without adaptation and with adaptation algorithms.

As one can easily see from presented results, the fuzzy algorithm GMDH shows better forecasting accuracy than classic GMDH for all adaptation algorithms.

So the results of experiments have confirmed indisputable advantages of fuzzy GMDH over classic GMDH for problem of forecasting macroeconomic indicators. In the next experiments, the comparison of fuzzy GMDH with results of neural network (NN) backpropagation was performed. The final results—MSE values on five forecasting points while forecasting CPI and GNP—are presented in **Table 1**.

Summing the experimental results, the following conclusions were made:

1. Forecasting accuracy of fuzzy GMDH algorithms are, in a whole, better than of non-fuzzy GMDH.

2. Forecasting accuracy of non-fuzzy and fuzzy GMDH algorithms are better than that of NN backpropagation. Modification of membership functions

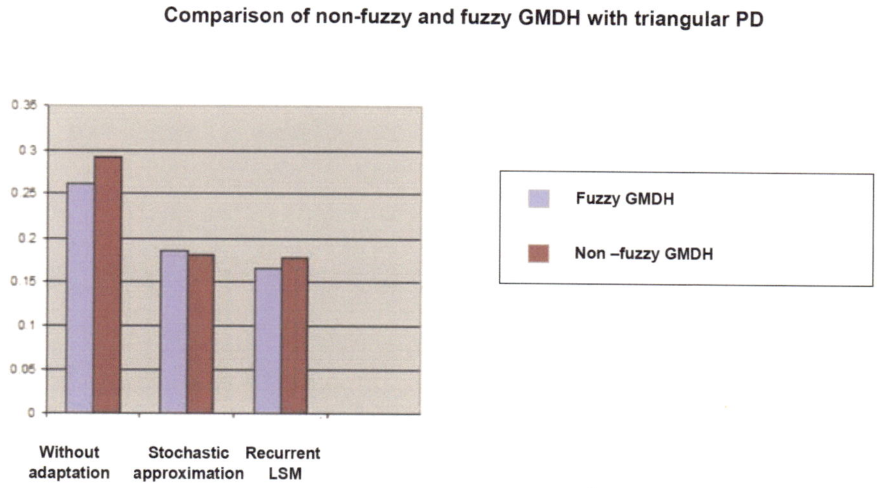

Figure 4.
Forecasting accuracy of classical and fuzzy GMDH for PCI.

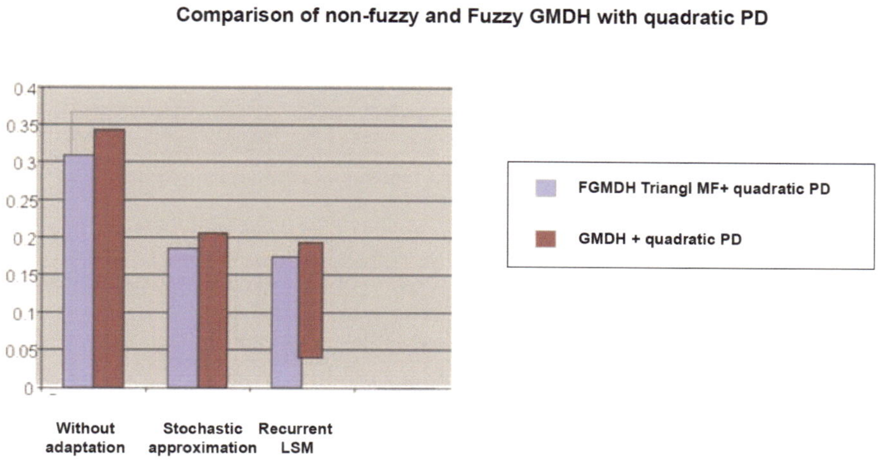

Figure 5.
Forecasting accuracy of classical and fuzzy GMDH for GDP.

doesn't lead to significant changes of forecasting quality, but the best results were obtained with bell-wise and Gaussian MF.

3. The best forecasting accuracy for considered problems was obtained with models of fuzzy GMDH using quadratic and trigonometric partial descriptions.

4. The best adaptation algorithm for fuzzy GMDH models is recurrent RLSM.

In [7, 8] the generalization of fuzzy GMDH for case when input data are also fuzzy was considered. Then a linear interval regression model takes the following form:

$$Y = A_0 Z_0 + A_1 Z_1 + \ldots + A_n Z_n,$$

Adaptation algorithm	Without adaptation		Stochastic approximation		RLSM	
	CPI	GNP	CPI	GNP	CPI	GNP
Triangle MF + quadratic polynomial	0.308	530.3	0.184	330.0	0.173	311.9
Gaussian MF + quadratic polynomial	0.294	531.3	—	—	—	—
Bell-wise MF + quadratic polynomial	0.268	497.9	—	—	—	—
Triangular MF + Chebyshev's polynomial	0.403	621.4	0.341	458.1	0.337	377. 2
Triangular MF + Laguerre polynomial	0.372	589.5	0.264	442.9	0.293	378.5
Triangular MF + trigonometric polynomial	0.261	537.7	0.185	347.9	0.165	331.9
Triangular MF + ARIMA model	0.862	704.3	0.683	513.5	0.597	472.6
GMDH + quadratic polynomial	0.343	596.7	0.204	428.2	0.192	369.2
GMDH + Chebyshev's polynomial	0.425	641.4	0.351	473.2	0.347	398.4
GMDH + Laguerre polynomial	0.396	598.5	0.292	459.0	0.274	376.4
GMDH + trigonometric polynomial	0.291	574.8	0.182	349.5	0.177	332.2
GMDH + ARIMA model	0.902	728.4	0.749	518.7	0.714	498.3
NN backpropagation[1]	0.954	792.3	—	—	—	—
NN backpropagation[2]	0.741	668.6	—	—	—	—

[1]*Neural network constructed with Neural Networks Toolbox 4.0.6 (MathWorks).*
[2]*Neural network constructed with Alyuda Forecaster 1.6 (Alyuda Research).*

Table 1.
Forecasting accuracy (MSE) for different forecasting methods.

Consider the case of symmetrical membership function for parameters Ai, so they can be described by the pair of parameters (a_i, c_i), where

$\underline{A}_i = a_i - c_i$, $\overline{A}_i = a_i + c_i$, c_i is the interval width, $c_i \geq 0$,

and Z_i is input variable which is also a fuzzy number of triangular shape, defined by three parameters $\left(\underline{Z_i}, \breve{Z}_i, \overline{Z_i}\right)$, where $\underline{Z_i}$ is a lower border, $\breve{Z}_i$ is a center, and $\overline{Z_i}$ is an upper border of fuzzy number.

It was shown that corresponding model is also LP problem, and corresponding algorithm FGMDH was developed for such case [7, 8].

8. Conclusions

In this paper fuzzy inductive modeling method FGMDH is considered.

The algorithms of FGMDH with different membership functions and different partial descriptions, including orthogonal polynomials, were presented and analyzed.

The experimental investigations of GMDH and fuzzy GMDH in problems of macroeconomic index forecast in Ukrainian economy were carried out.

The comparative investigations of FGMDH with ARIMA and neural network backpropagation were performed.

Experimental result analysis has confirmed the high accuracy of fuzzy GMDH in problems of forecasting in macroeconomy.

Author details

Yu. Zaychenko* and Helen Zaychenko
Igor Sikorsky Kyiv Polytechnic Institute, Kiev, Ukraine

*Address all correspondence to: zaychenkoyuri@ukr.net

References

[1] Ivakhnenko AG, Mueller IA. Self-Organization of Forecasting Models. Kiev: Publ. House "Technika"; 1985

[2] Ivakhnenko AG, Zaychenko YP, Dimitrov VD. Decision-Making on the Basis of Self-Organization. Moscow: Publ. house "Soviet Radio"; 1976. p. 363

[3] Zgurovsky Mikhail Z, Zaychenko YP. Inductive modeling method (GMDH) in problems of intellectual data analysis and forecasting. In: Studies in Computational Intelligence. Switzerland: Springer Intern, Publishing AG; 2016. p. 406

[4] Zaychenko YP, Kebkal AG, Krachkovsky VF. Fuzzy group method of data handling and its application for macro-economic indicators forecasting. Scientific Papers of NTUU "KPI". 2000; **2**:18-26

[5] Zaychenko YP, Zayets IОЮП. Synthesis and adaptation of fuzzy forecasting models on the basis of self-organization method. Scientific Papers of NTUU "KPI". 2001;**3**:34-41

[6] Zaychenko Y, Zayets IO, Kamotsky OV, Pavlyuk OV. The investigations of different membership functions in Fuzzy Group Method of Data Handling. Control Systems and machines. 2003;**2**: 56-67

[7] Zaychenko YP. Fuzzy models and methods in intellectual systems. Kiev: Publ. House "Slovo"; 2008. p. 354

[8] Zaychenko Y. Fuzzy Group Method of Data Handling under fuzzy input data. System Research and Information Technologies. 2007;**3**:100-112

Section 4

Applications

Chapter 8

Segmenting Images Using Hybridization of K-Means and Fuzzy C-Means Algorithms

Raja Kishor Duggirala

Abstract

Image segmentation is an essential technique of image processing for analyzing an image by partitioning it into non-overlapped regions each region referring to a set of pixels. Image segmentation approaches can be divided into four categories. They are thresholding, edge detection, region extraction and clustering. Clustering techniques can be used for partitioning datasets into groups according to the homogeneity of data points. The present research work proposes two algorithms involving hybridization of K-Means (*KM*) and Fuzzy C-Means (*FCM*) techniques as an attempt to achieve better clustering results. Along with the proposed hybrid algorithms, the present work also experiments with the standard K-Means and *FCM* algorithms. All the algorithms are experimented on four images. CPU Time, clustering fitness and sum of squared errors (SSE) are computed for measuring clustering performance of the algorithms. In all the experiments it is observed that the proposed hybrid algorithm *KMandFCM* is consistently producing better clustering results.

Keywords: image segmentation, clustering, K-Means, Fuzzy C-Means, hybridization, sum of squared error, clustering fitness

1. Introduction

Images are often the most important category among the available digital data. In the recent years, image data is increasing and will continue increase in the near future. Since it is difficult to deal with large amount of image data as the available data increases, it becomes crucial to use the automated tools for various purposes in connection to image data. The image processing provides wide range of techniques to deal with the images. By using the image processing techniques, we can make the work much easier not only for now, but also for the future when there will be more data and more work to do on the images.

Image segmentation is an essential image processing technique that analyzes an image by partitioning it into non-overlapped regions each region referring to a set of pixels. The pixels in a region are similar with respect to some characteristic such as color, intensity, or texture [1]. The pixels significantly differ with those in the other regions with respect to the same characteristic [2–4]. Image segmentation plays an important role in a variety of applications such as robot vision, object recognition, medical imaging and etc. [5–7]. Image segmentation approaches can be divided into four categories. They are thresholding, edge detection, region extraction and clustering. Clustering techniques can be used for data segmenting image data as they are used for partitioning large datasets into groups according to the homogeneity of data points.

In clustering, a given population of data is partitioned into groups such that objects are similar to one another within the same group and are dissimilar to the objects in other groups [8, 9]. There are different categories of clustering techniques. These can be partitional (hierarchical and non-hierarchical), like K-means, PAM, CLARA, CLARANs [10, 11]; model-based, like Expectation Maximization, SOM, Mixture model clustering [12, 13]; or fuzzy-based like Fuzzy C-Means [14, 15].

Partitional clustering techniques attempt to break a population of data into some predefined number of clusters such that the partition optimizes a given criterion.

Formally, clusters can be seen as subsets of the given dataset. So, clustering methods can be classified according to whether the subsets are fuzzy or crisp (hard). In hard clustering, an object either does or does not belong to a cluster. These methods partition the data into a specified number of mutually exclusive subsets. However, in fuzzy-based clustering, the objects may belong to several clusters with different degrees of membership [16].

It is studied in the literature that many researchers experimented with the Fuzzy C-Means (*FCM*) algorithm in a wide variety of ways for achieving better image segmentation results [1, 17]. In [18], a penalized *FCM* (*PFCM*) algorithm is presented for image segmentation for handling noise by adjusting a penalty coefficient. The penalty term used here takes the spatial dependence of the objects into consideration, which is modified according to the *FCM* criterion. In [19], a fuzzy rule-based technique is proposed. It employs the rule-based neighborhood enhancement system to impose spatial continuity by post-processing on the clustering results obtained using *FCM* algorithm. In [20], a Geometrically Guided *FCM* (*GG-FCM*) algorithm is proposed, which is based on a semi-supervised *FCM* technique for multivariate image segmentation. In [21], a regularization term was introduced into the standard *FCM* to impose the neighborhood effect. In [22], this regularization term is incorporated into a kernel-based fuzzy clustering algorithm. In [23], this regularization) term is incorporated into the adaptive *FCM* (*AFCM*) algorithm [24] to overcome the noise sensitivity of *AFCM* algorithm.

However, it is found in the literature that a very less attention is paid towards the hybridization of clustering techniques for partitioning the datasets.

The present research work aims at developing hybrid clustering algorithms involving K-Means and Fuzzy C-Means (*FCM*) techniques for achieving better clustering results. As part of hybridization, two algorithms are developed, *KMFCM* and *KMandFCM*. The *KMFCM* algorithm first performs K-Means on the dataset and then performs *FCM* using the results of K-Means. The *KMandFCM* algorithm performs K-Means and *FCM* in the alternative iterations.

All the experiments are carried out using the datasets that are related to four images. For performance evaluation, CPU time, clustering fitness and sum of squared error (SSE) are taken into consideration.

The following sections provide a detailed discussion of K-Means (*KM*), Fuzzy C-Means (*FCM*), *KMFCM* and *KMandFCM* algorithms.

2. The K-Means (*KM*) algorithms

Partitional clustering methods are appropriate for the efficient representation of large datasets [11]. These methods determine k clusters such that the data objects in a cluster are more similar to each other than to the objects in other clusters.

The K-Means is a partitional clustering method, which partitions a given dataset into a pre-specified number, k, of clusters [25]. It is a simple iterative method. The algorithm is initialized by randomly choosing k points from a given dataset as the initial cluster centers, i.e., cluster means. The algorithm iterates through two steps till its convergence:

1. Data assignment: this step partitions the data by assigning each data object to its closest cluster center.

2. Updating the cluster centers: update the center of each cluster based on the objects assigned to that cluster.

The algorithm for K-Means is as follows [26]. Here, k represents the number of clusters, d represents the number of dimensions or attributes, X_i represents the ith data sample, μ_j (j = 1, 2, ..., k) represents the mean vector of cluster C_j, t is the iteration number. For termination condition the algorithm computes *percentage change*, Eq. (2). The algorithm terminates when *Percentage change* $< \alpha$. Here, α is assumed to be 3 since it is negligible.

***KM* algorithm**

1. Select k vectors randomly from the dataset as the initial cluster centers, μ_j (j = 1, 2, ..., k). Set the current iteration $t = 0$.

2. Assign each vector, X_i, to its closest cluster center using Euclidean distance, Eq. (1).

$$d\left(X_i, \mu_j\right) = \sqrt{\sum_{l=1}^{d}\left(x_{il} - \mu_{jl}\right)^2} \tag{1}$$

3. Update mean vectors μ_j (j = 1, ..., k).

4. Compute Percentage change as follows

$$\text{Percentage change} = \frac{|\Psi_t - \Psi_{t+1}|}{\Psi_t} \times 100 \tag{2}$$

where Ψ_t is the number of vectors assigned to new clusters in tth iteration and Ψ_{t+1} is the number of vectors assigned to new clusters in $(t + 1)$th iteration.

5. Stop the process if *Percentage change* $< \alpha$, otherwise set $t = t + 1$ and repeat the steps 2–4 with the updated parameter.

The K-Means uses Euclidean distance as a proximity measure for determining the closest cluster to which a data object is assigned [13]. The algorithm stops when the assignment of data points to the clusters no longer changes or some other criterion is satisfied. The K-Means is a widely used algorithm for clustering and it requires less CPU time. However, it mainly suffers from detecting the natural clusters that have non-spherical shapes or widely different sizes or densities [25].

3. The Fuzzy C-Means (*FCM*) algorithms

Fuzzy-based clustering techniques focus on modeling uncertain and vague information that is found in the real world situations. These techniques deal with the clusters whose boundaries cannot be defined sharply [14, 15]. By fuzzy-based clustering, one can know if data objects fully or partially belong to the clusters based on their memberships in different clusters [27]. Among the fuzzy-based clustering methods, Fuzzy C-Means (*FCM*) is the most well-known algorithm as it has the advantage of robustness for obscure information about the clusters [1, 28].

In *FCM*, a dataset is grouped into k clusters, where every data object may relate to every cluster with some degree of membership to that cluster [16]. The membership of a data object towards a cluster can range between 0 and 1 [29]. The sum of memberships for each data point must be unity.

The *FCM* iterates through two phases for converging to a solution. First, each data object will be associated with a membership value for each cluster, and second, assigning the data object to the cluster with the highest membership value [2].

The algorithm for *FCM* is given below [30]. Here, U is the $k \times N$ membership matrix. While computing the cluster centers and updating the membership matrix at each iteration, the *FCM* uses membership weight, m. For most data $1.5 \leq m \leq 3.0$ gives good results [29]. In all our experiments, we take $m = 1.25$.

***FCM* algorithm**

1.Initialize parameters: select k vectors randomly as cluster means; set initial membership matrix $U^{(0)}_{k \text{ X } N}$, set the current iteration $t = 0$.

2.Assign each data object X_i to clusters using the membership matrix.

3.Compute jth cluster center as follows:

$$\mu_j^{t+1} = \frac{\sum_{i=1}^{N} (u_{ji})^m X_i}{\sum_{i=1}^{N} (u_{ji})^m} \tag{3}$$

4.Compute new membership matrix using

$$u_{ji}^{t+1} = \left[\sum_{l=1}^{k} \left(\frac{\left\| X_i - \mu_j^{t} \right\|^2}{\left\| X_i - \mu_l^{t} \right\|^2} \right)^{1/m - 1} \right]^{-1} \tag{4}$$

5.Assign each data object X_i to clusters using the membership matrix.

6.Compute *Percentage change* using Eq. (2).

7.Stop the process if the *Percentage change* is $<\alpha$. Otherwise, set $t = t + 1$ and repeat the steps 3–7 with the updated parameters.

FCM is widely studied and applied in geological shape analysis [31], medical diagnosis [32], automatic target recognition [33], meteorological data [28], pattern recognition, image analysis, image segmentation and image clustering [34–36], agricultural engineering, astronomy, chemistry [37], detection of polluted sites [38] and etc.

4. Hybridization involving K-Means and *FCM* techniques

The partitional [11] and fuzzy-based [16] methods are widely applied clustering techniques in several areas. The partitional clustering methods do hard clustering, where the dataset is partitioned into a specified number of mutually exclusive subsets. The K-Means, as a partitional clustering method is found in the research literature as widely applied technique in a variety of experiments. While clustering the data, the K-Means aims at minimizing the local distortion [39, 40]. However, K-Means is ideal if the data objects are distributed in well-separated groups.

In fuzzy-based clustering, objects are not forced to fully belong to one cluster. Here, an object may belong to many clusters with varying degrees of membership. This membership can range between 0 and 1 indicating the partial belongingness of objects to the clusters [16]. Fuzzy clustering techniques help in understanding if the data objects fully or partially belong to clusters depending on their memberships [27]. In *FCM*, each data object belongs to each cluster with some degree of membership that ranges between 0 and 1 [29]. Here, clusters are treated as fuzzy sets. In general, introducing the fuzzy logic in K-Means is the Fuzzy C-Means algorithm [41].

The following sub-section discusses two algorithms that apply hybridization of K-Means (*KM*) and Fuzzy C-Means (*FCM*) clustering techniques [42]. These algorithms are *KMFCM* and *KMandFCM*. The *KMFCM* algorithm first performs K-Means on the given dataset and then performs the *FCM* using the results of K-Means. The *KMandFCM* algorithm performs K-Means and *FCM* in the alternative iterations on the given dataset. The detailed discussion of these hybrid algorithms is presented in the following subsections.

4.1 The *KMFCM* algorithm

The proposed hybrid clustering algorithm *KMFCM* first performs the K-Means (*KM*) technique completely on the given dataset. Using the resulted cluster centers of *KM* as cluster seeds, the *FCM* is performed on the given dataset till termination. Here, to run the first iteration of the *FCM*, the cluster centers and the membership matrix are calculated based on the results of *KM*. The remaining iterations continue as in the *FCM* algorithm.

The algorithm for the *KMFCM* is given below. Here, *KM*-Step is the K-Means step and *FCM*-Step is the Fuzzy C-Means step.

***KMFCM* algorithm**

1. ***KM*-Step**: select k vectors randomly from the dataset as the initial cluster centers μ_j ($j = 1, \ldots, k$). Set the current iteration $t = 0$.
2. Assign each data object X_i to its closest cluster center using Eq. (1).
3. Update cluster centers μ_j ($j = 1, \ldots, k$) and set $t = t + 1$.
4. Compute *Percentage change* using Eq. (2).
5. If *Percentage change* $\geq \alpha$, repeat steps 2–4.
6. ***FCM*-Step**: compute the membership matrix $U^{(t)}_{k \times N}$ using Eq. (4) based on the results of *KM*-Step.
7. Assign data objects to clusters using membership matrix.
8. For each cluster C_j, compute the center μ_j ($j = 1, \ldots, k$) using Eq. (3)
9. Compute *Percentage change* using Eq. (2).
10. Stop the process if *Percentage change* $< \alpha$. Otherwise, set $t = t + 1$ and repeat steps 6–9.

4.2 The *KMandFCM* algorithm

Clustering in *KMandFCM* is performed by executing K-Means and the *FCM* techniques in alternative iterations on the given dataset till termination. The first iteration is performed using K-Means assuming some randomly selected data points as cluster centers. The second iteration is performed using *FCM* technique. For this iteration the cluster means, covariance matrices and the membership matrix are calculated using the results of first iteration. Third iteration is performed using K-Means technique. This iteration computes cluster means using results obtained from the second iteration. In this way, clustering is performed using K-Means and *FCM* in the alternative iterations till termination.

The algorithm for the proposed *KMandFCM* algorithm is given below. Here, *KM*-Step is the K-Means step and *FCM*-Step is the Fuzzy C-Means step.

***KM and FCM* algorithm**

1.Select k vectors randomly from the dataset as initial cluster centers μ_j ($j = 1, \ldots, k$). Set the current iteration $t = 0$.

2.***KM*-Step**: assign each vector X_i to its closest cluster center using Eq. (1).

3.***FCM*-Step**: set $t = t + 1$.

4.For each cluster C_j, compute the center μ_j using Eq. (3)

5.Compute the new membership matrix $U^{(t)}_{k \text{ X } N}$ using Eq. (4)

6.Assign data objects to clusters using the membership matrix.

7.Compute *Percentage change* using Eq. (2).

8.Stop the process if *Percentage change* $< \alpha$. Otherwise, set $t = t + 1$.

9.***KM*-Step**: For each cluster C_j, compute new center μ_j using Eq. (3).

10.Assign each vector X_i to its closest cluster center using Eq. (1).

11.Compute *Percentage change* using Eq. (2).

12.Stop the process if *Percentage change* $< \alpha$. Otherwise, go to step 3.

For all the algorithms, i.e., *KM*, *FCM*, *KMFCM*, *KMandFCM*, the same termination condition, Eq. (2), is used.

5. Performance evaluation measures

For performance evaluation of algorithms, CPU time in seconds, sum of squared error [12] and clustering fitness [43] are taken into consideration and are calculated for all the algorithms.

5.1 Sum of squared errors

The objective of clustering is to minimize the within-cluster sum of squared error (SSE). The lesser the SSE, the better the goodness of fit is. The sum of squared error [12] for the results of each clustering algorithm is computed using the Eq. (5)

$$SSE = \sum_{j=1}^{k} \sum_{X_i \in C_j} \left(X_i - \mu_j\right)^2 \qquad (5)$$

Here, X_i is the ith data object in the dataset, μ_j (j = 1, ..., k) is the center of the cluster C_j, and k is the number of clusters.

5.2 Clustering fitness

The main objective of any clustering algorithm is to generate clusters with higher intra-cluster similarity and lower inter-cluster similarity. So, it is also important to consider inter-cluster similarity while evaluating the clustering performance. In the present work, clustering fitness is also considered as a performance criterion, which requires the calculation of both intra-cluster similarity and inter-cluster similarity. The computation of clustering fitness also requires the experiential knowledge, λ. The computation of clustering fitness results in higher value when the inter-cluster similarity is low and results in lower value for when the inter-cluster similarity is high. Also that to make the computation of clustering fitness unbiased, the value of λ is taken as 0.5 [43].

(a) **Intra-cluster similarity for the cluster C_j**: it can be quantified via a function of the reciprocals of intra-cluster radii within each of the resulting clusters. The intra-cluster similarity [43] of a cluster C_j ($1 = j = k$), denoted as $S_{tra}(C_j)$, is defined in Eq. (6)

$$S_{tra}\left(C_j\right) = \frac{1+n}{1+\sum_1^n dist(I_l, Centroid)} \qquad (6)$$

Here, n is the number of items in cluster C_j, I_j ($1 = j = n$) is the jth item in cluster C_j, and dist(I_j, $Centroid$) calculates the distance between I_j and the centroid of C_j, which is the intra-cluster radius of C_j. To smooth the value of $S_{tra}(C_j)$ and allow for possible singleton clusters, 1 is added to the denominator and numerator.

(b) **Intra-cluster similarity for one clustering result C**: it is denoted as $S_{tra}(C)$. It is defined in Eq. (7), [43]

$$S_{tra}(C) = \frac{\sum_1^k S_{tra}\left(C_j\right)}{k} \qquad (7)$$

Here, k is the number of resulting clusters in C and $S_{tra}(C_j)$ is the intra-cluster similarity for the cluster C_j.

(c) **Inter-cluster similarity**: it can be quantified via a function of the reciprocals of inter-cluster radii of the clustering centroids. The inter-cluster similarity [43] for one of the possible clustering results C, denoted as $S_{ter}(C_j)$, is defined as Eq. (8)

$$S_{ter}(C) = \frac{1+k}{1+\sum_1^k dist\left(Centroid_j, Centroid^2\right)} \qquad (8)$$

Here, k is the number of resulting clusters in C, $1 = j = k$, $Centroid_j$ is the centroid of the jth cluster in C, $Centroid^2$ is the centroid of all centroids of clusters in C. We compute inter-cluster radius of $Centroid_j$ by calculating dist($Centroid_j$, $Centroid^2$), which is distance between $Centroid_j$, and $Centroid^2$. To smooth the value of $S_{ter}(C)$

and allow for possible all-inclusive clustering result, 1 is added to the denominator and the numerator.

(d) Clustering fitness: the clustering fitness [43] for one of the possible clustering results C, denoted as CF, is defined as Eq. (9)

$$CF = \lambda \times S_{tra}(C) + \frac{1-\lambda}{S_{ter}(C)} \tag{9}$$

Here, λ $(0 < \lambda < 1)$ is an experiential weight, $S_{tra}(C)$ is the intra-cluster similarity for the clustering result C and $S_{ter}(C)$ is the inter-cluster similarity for the clustering result C. To avoid biasedness in our experiments, λ is assumed to be 0.5.

6. Experiments and results

Experimental work has been carried out on the system with Intel(R) Core(TM) i3-5005U CPU@2.00GHz processor speed, 4GB RAM, Windows 7 OS (64-bit) and using JDK1.7.0_45. Separate modules are written for each of the above discussed methods to observe the CPU time for clustering any dataset by keeping the cluster seeds same for all methods. I/O operations are eliminated and the CPU time observed is strictly for clustering of the data.

Along with the newly developed hybrid algorithms, experiments are also conducted with the algorithms for standard K-Means (*KM*) and Fuzzy C-Means (*FCM*) for performance comparison. All the algorithms are executed using datasets that are related to four images. The details of these images are available in **Table 1**.

SNO	Image	Resolution	No. of points	No. of dimensions
1	Heart	341 × 367	125,147	3
2	Kidneys	473 × 355	167,915	3
3	Baboon	512 × 512	262,144	3
4	Lena	256 × 256	65,536	3

Table 1.
Medical Images.

The medical images used in the present experiment are heart image [44] and kidneys image [45] (**Figures 1** and **2**). The experiments are also carried out using two benchmark images. They are Baboon and Lena images [46] (**Figures 3** and **4**).

Below is the brief description of medical images.

The Heart is a medical image obtained from biology data repository [44]. It is in "jpeg" format. The 'Kidneys' is a colored MRI scan of a coronal section through a human abdomen, showing the front view of healthy kidneys and liver [45]. It is in 'jpeg' format. The Baboon and Lena are benchmark test images that are found frequently in the literature [46]. These are all in uncompressed "tif" format.

All the algorithms for standard K-Means (*KM*), standard Fuzzy C-Means (*FCM*), *KMFCM* and *KMandFCM* are executed on each image data with varying number of clusters (k = 10, 11, 12, 13, 14, 15). For all algorithms, same cluster seeds are used. Same termination condition Eq. (2) is used for all the experiments. The details of CPU time, clustering fitness and SSE of each algorithm for the all images are given in the following sub-sections (**Tables 2–13**). The results are also projected in their respective graphs (**Figures 5–16**).

6.1 Observations with Heart image

K	*KM*	*FCM*	*KMFCM*	*KM and FCM*
10	0.21	0.30	1.36	0.19
11	0.21	0.32	1.48	0.20
12	0.25	0.40	1.61	0.20
13	0.09	0.35	1.58	0.22
14	0.14	0.39	1.73	0.23
15	0.36	0.43	2.15	0.26

Table 2.
CPU time of each clustering technique (Heart image).

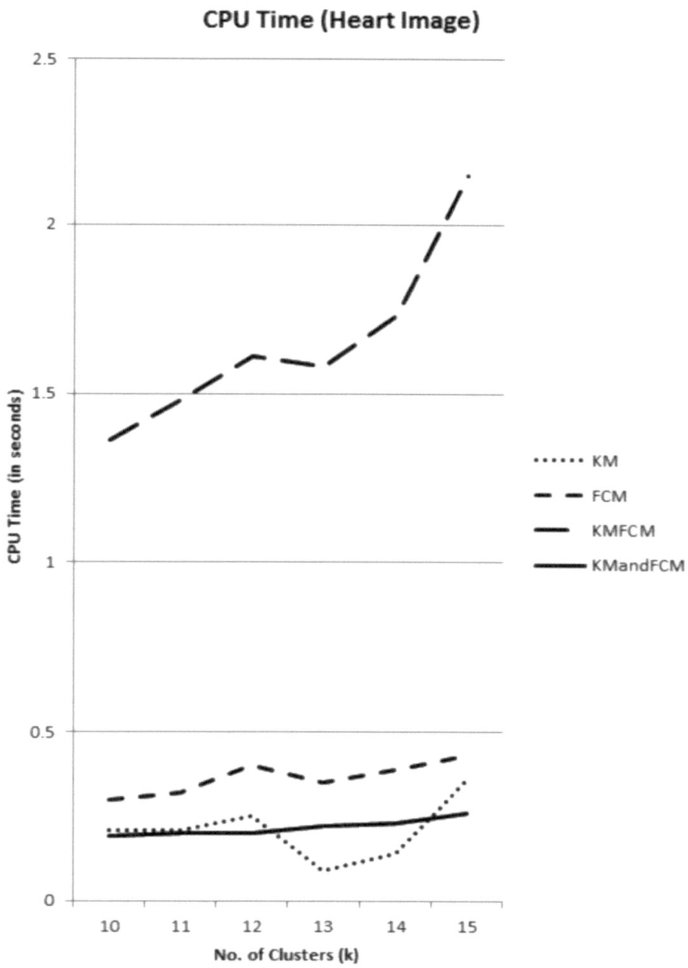

Figure 1.
CPU time (Heart image).

K	*KM*	*FCM*	*KMFCM*	*KM and FCM*
10	51.20	56.62	58.51	64.78
11	49.79	55.73	55.40	62.14
12	42.27	55.80	61.16	65.97
13	34.88	47.54	41.08	58.46
14	48.34	55.22	56.62	60.35
15	47.54	57.96	48.24	59.22

Table 3.
Clustering fitness of each clustering technique (Heart image).

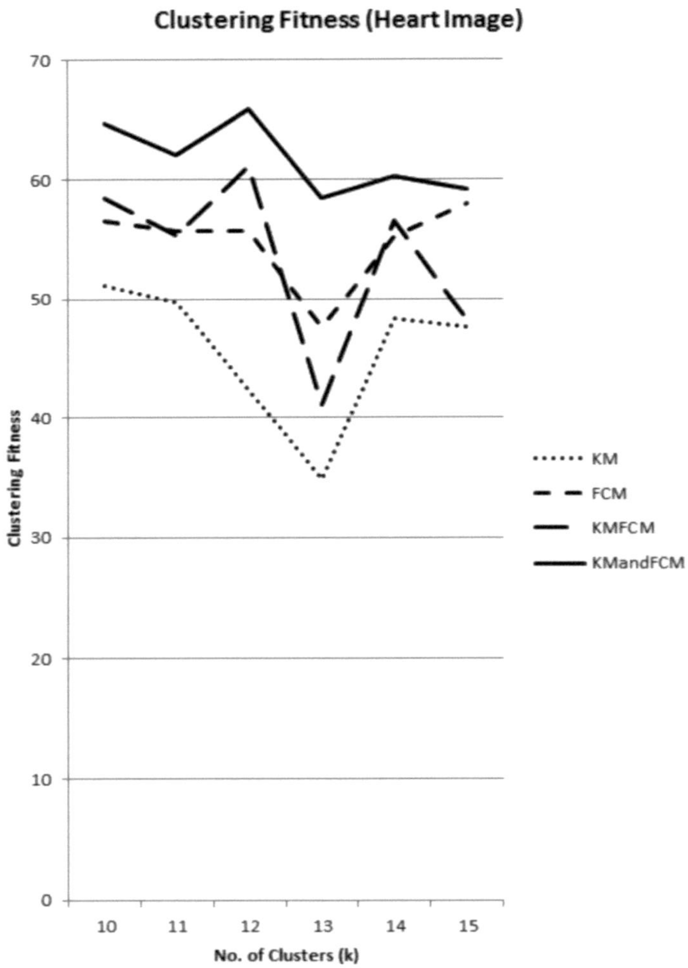

Figure 2.
Clustering Fitness (Heart image).

K	*KM*	*FCM*	*KMFCM*	*KM and FCM*
10	0.0163	0.0152	0.0148	0.0041
11	0.0150	0.0145	0.0074	0.0036
12	0.0173	0.0163	0.0059	0.0031
13	0.0185	0.0171	0.0285	0.0037
14	0.0142	0.0139	0.0113	0.0028
15	0.0138	0.0114	0.0241	0.0024

Table 4.
SSE of each clustering technique (Heart image).

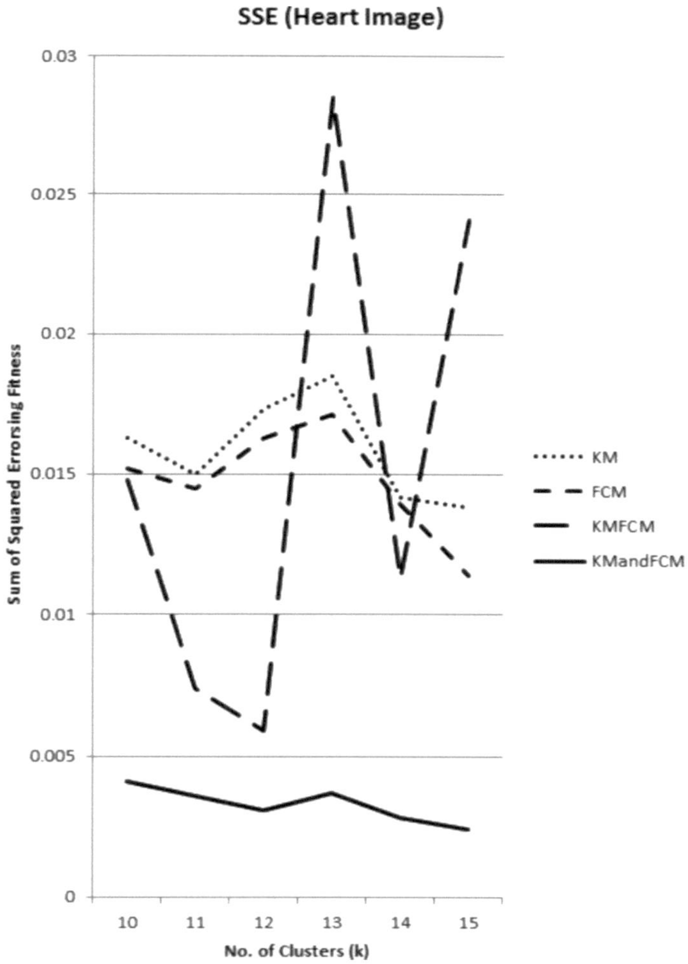

Figure 3.
Sum of squared errors (Heart image).

6.2 Observations with Kidneys image

K	*KM*	*FCM*	*KMFCM*	*KM and FCM*
10	0.09	0.68	1.58	0.55
11	0.13	0.41	1.83	0.26
12	0.81	0.58	2.64	0.46
13	0.08	0.47	2.07	0.30
14	0.24	0.60	2.40	0.31
15	0.65	1.78	2.22	1.06

Table 5.
CPU time of each clustering technique (Kidneys image).

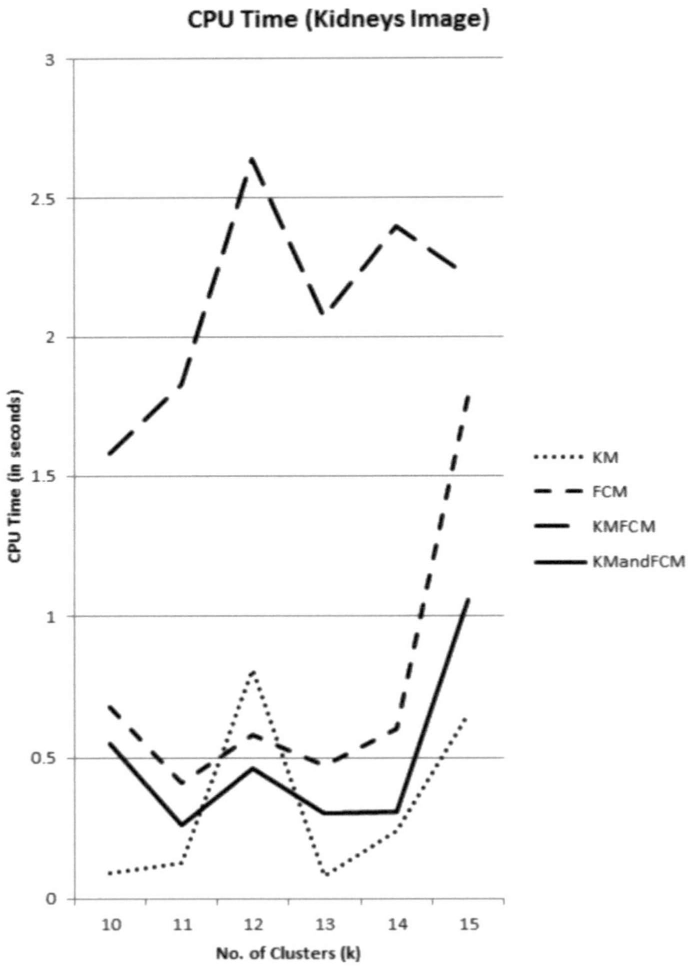

Figure 4.
CPU time (Kidneys image).

K	*KM*	*FCM*	*KMFCM*	*KM and FCM*
10	38.40	47.15	54.76	61.48
11	42.11	49.43	57.86	65.84
12	52.41	61.03	60.00	65.41
13	41.20	51.04	48.73	56.79
14	57.49	64.85	64.88	71.59
15	53.10	61.40	62.85	66.42

Table 6.
Clustering fitness of each clustering technique (Kidneys image).

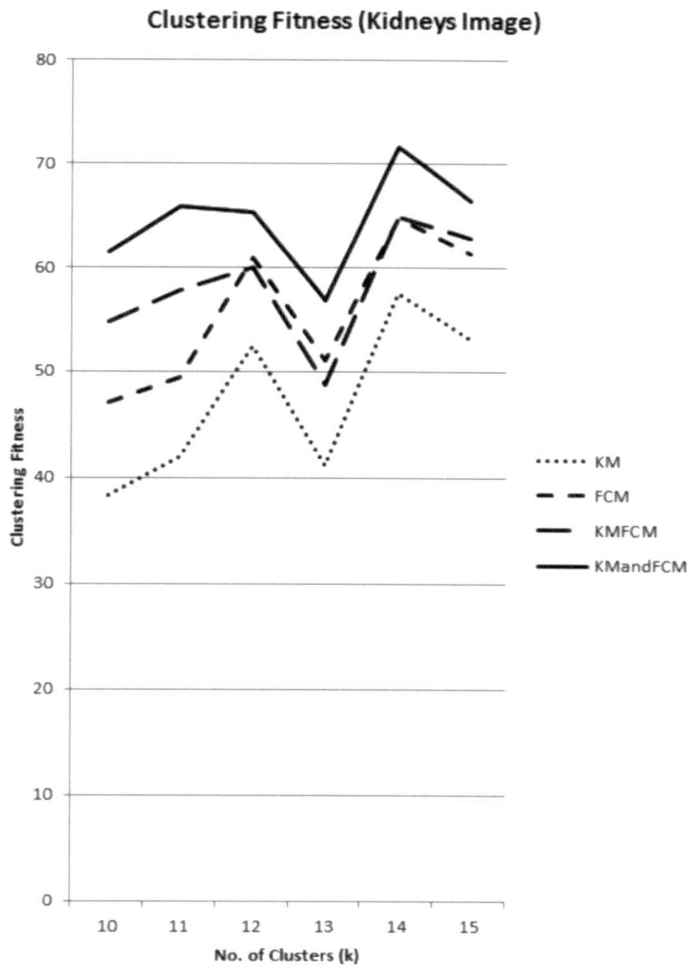

Figure 5.
Clustering Fitness (Kidneys image).

K	KM	FCM	KMFCM	KM and FCM
10	0.0281	0.0215	0.0129	0.0075
11	0.0265	0.0172	0.0114	0.0054
12	0.0249	0.0109	0.0140	0.0029
13	0.0123	0.0109	0.0191	0.0112
14	0.0144	0.0090	0.0067	0.0037
15	0.0115	0.0045	0.0028	0.0011

Table 7.
SSE of each clustering technique (Kidneys image).

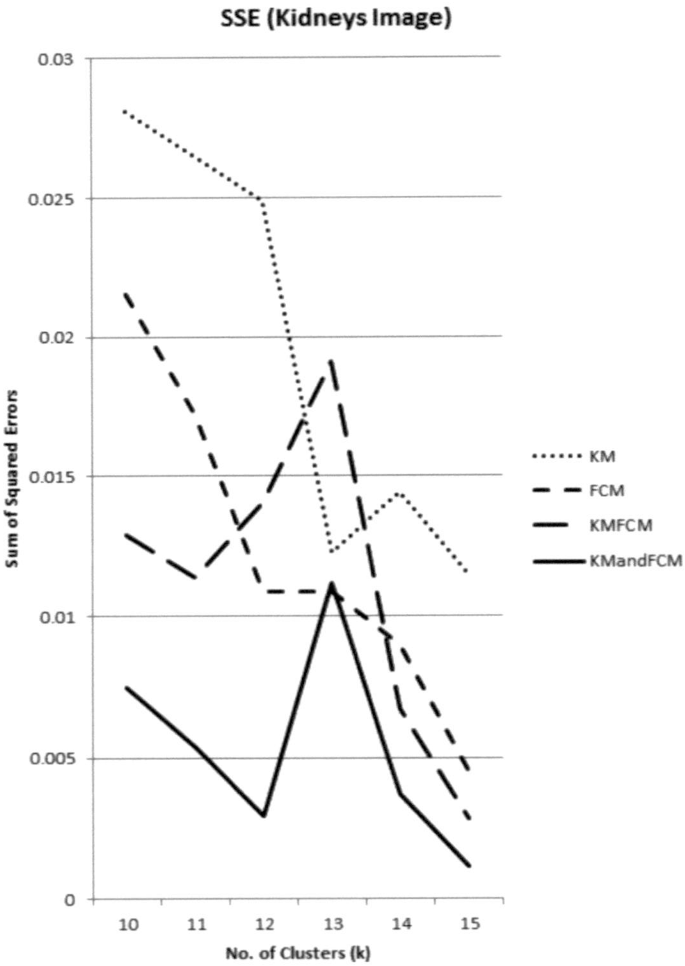

Figure 6.
Sum of squared errors (Kidneys image).

6.3 Observations with Baboon image

K	*KM*	*FCM*	*KMFCM*	*KM and FCM*
10	0.14	0.79	2.16	0.62
11	0.16	0.86	2.37	0.63
12	0.29	0.91	2.68	0.63
13	0.31	1.01	2.91	0.50
14	0.36	0.72	3.14	0.78
15	0.48	1.10	3.24	0.55

Table 8.
CPU time of each clustering method (Baboon image).

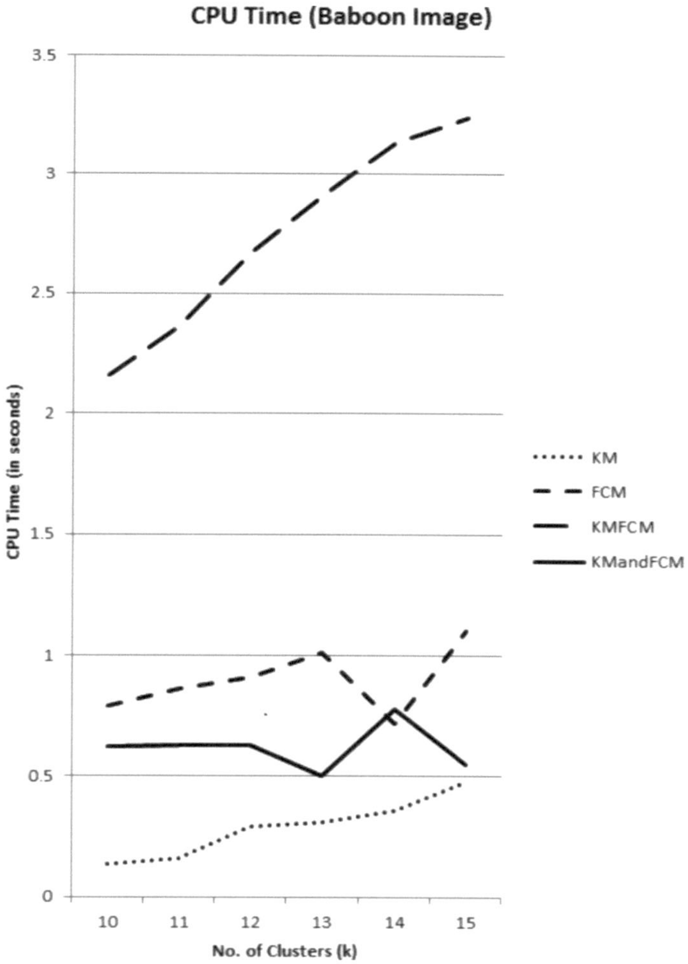

Figure 7.
CPU time (Baboon image).

K	*KM*	*FCM*	*KMFCM*	*KM and FCM*
10	30.22	32.17	36.02	39.07
11	22.28	29.71	37.36	39.49
12	28.70	32.63	35.13	39.57
13	31.28	33.47	40.39	42.28
14	25.92	29.49	37.77	39.81
15	36.48	38.16	34.43	39.98

Table 9.
Clustering fitness of each clustering method (Baboon image).

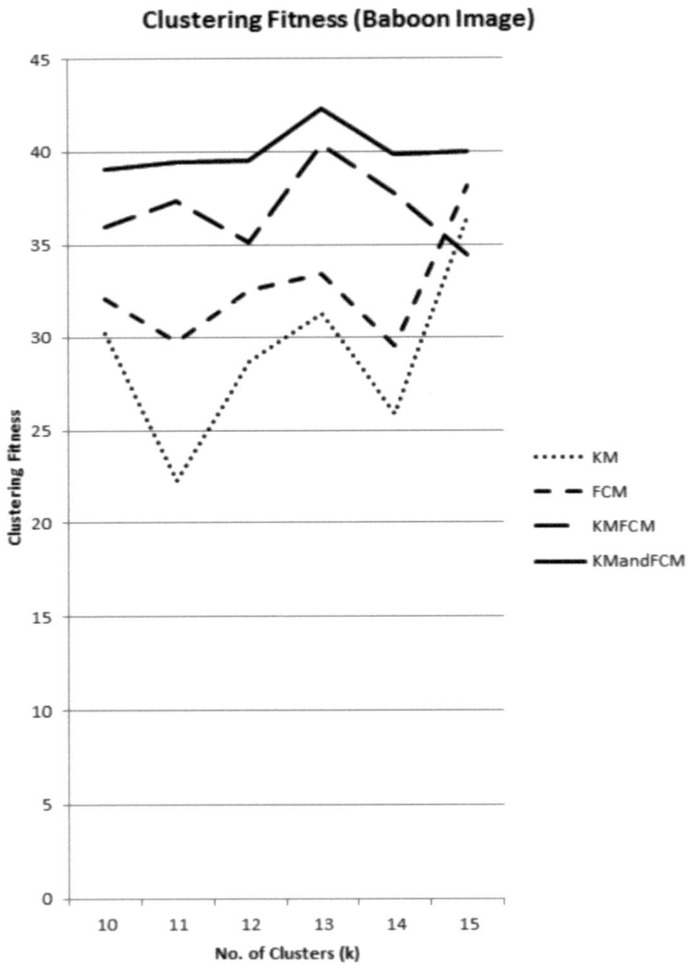

Figure 8.
Clustering fitness (Baboon image).

K	*KM*	*FCM*	*KMFCM*	*KM and FCM*
10	0.0080	0.0063	0.0059	0.0030
11	0.0073	0.0068	0.0037	0.0024
12	0.0099	0.0071	0.0053	0.0029
13	0.0065	0.0058	0.0070	0.0025
14	0.0087	0.0070	0.0041	0.0022
15	0.0069	0.0056	0.0027	0.0019

Table 10.
SSE of each clustering method (Baboon image).

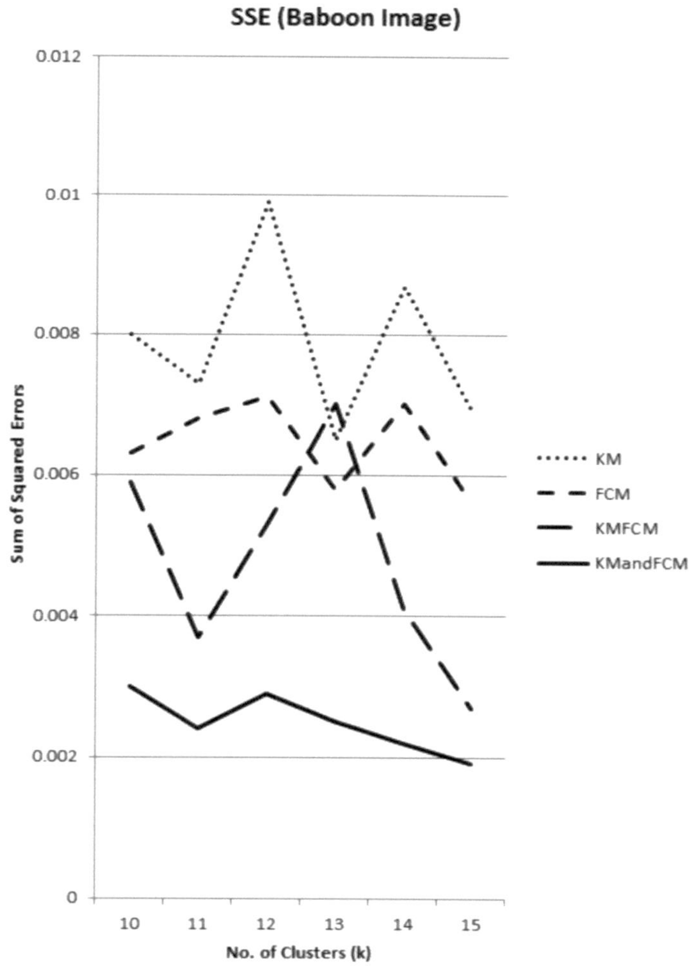

Figure 9.
Sum of squared errors (Baboon image).

6.4 Observations with Lena image

K	KM	FCM	KMFCM	KMandFCM
10	0.08	0.15	0.66	0.09
11	0.13	0.44	0.76	0.32
12	0.06	0.17	0.77	0.11
13	0.09	0.40	0.84	0.32
14	0.05	0.20	0.92	0.13
15	0.21	0.24	1.09	0.14

Table 11.
CPU time of each clustering method (Lena image).

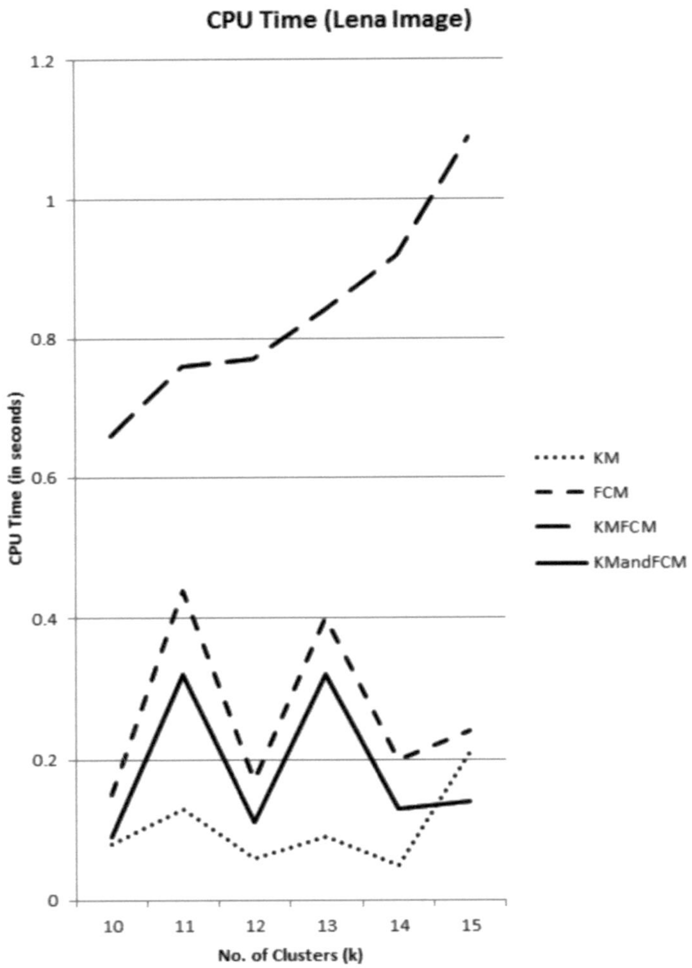

Figure 10.
CPU time (Lena image).

K	*KM*	*FCM*	*KMFCM*	*KM and FCM*
10	25.50	28.80	30.61	32.79
11	22.97	25.52	27.95	31.08
12	20.22	23.38	25.44	29.97
13	28.71	30.13	32.74	34.26
14	26.75	29.83	31.05	33.27
15	23.70	30.19	32.79	34.60

Table 12.
Clustering fitness of each clustering method (Lena image).

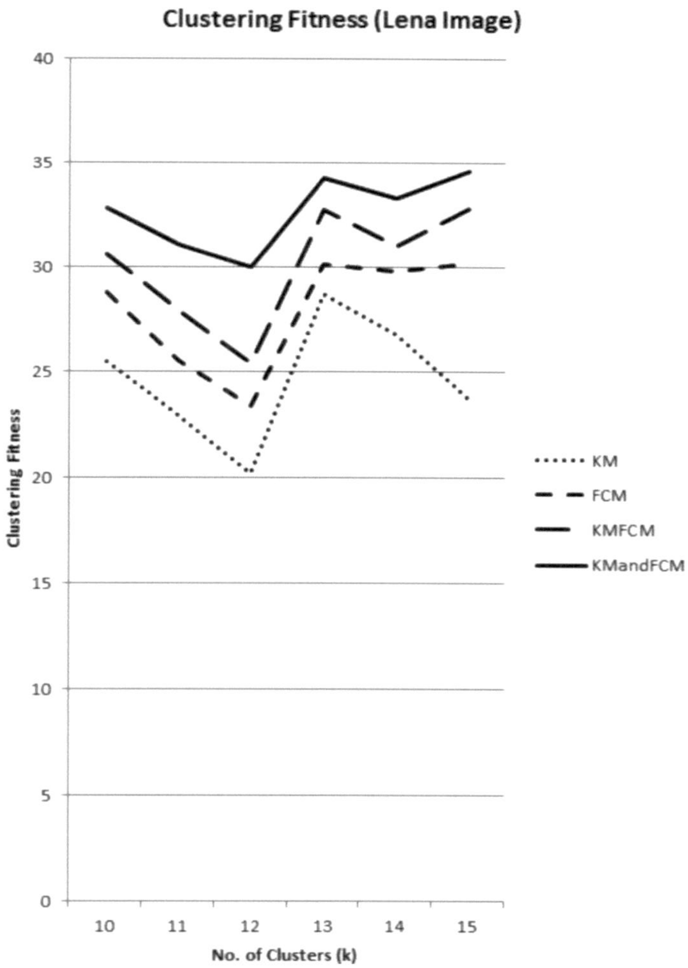

Figure 11.
Clustering fitness (Lena image).

K	KM	FCM	KMFCM	KM and FCM
10	0.0147	0.0127	0.0093	0.0034
11	0.0245	0.0218	0.0099	0.0041
12	0.0246	0.0178	0.0077	0.0034
13	0.0144	0.0106	0.0060	0.0027
14	0.0135	0.0110	0.0062	0.0024
15	0.0130	0.0100	0.0049	0.0022

Table 13.
SSE of each clustering method (Lena image).

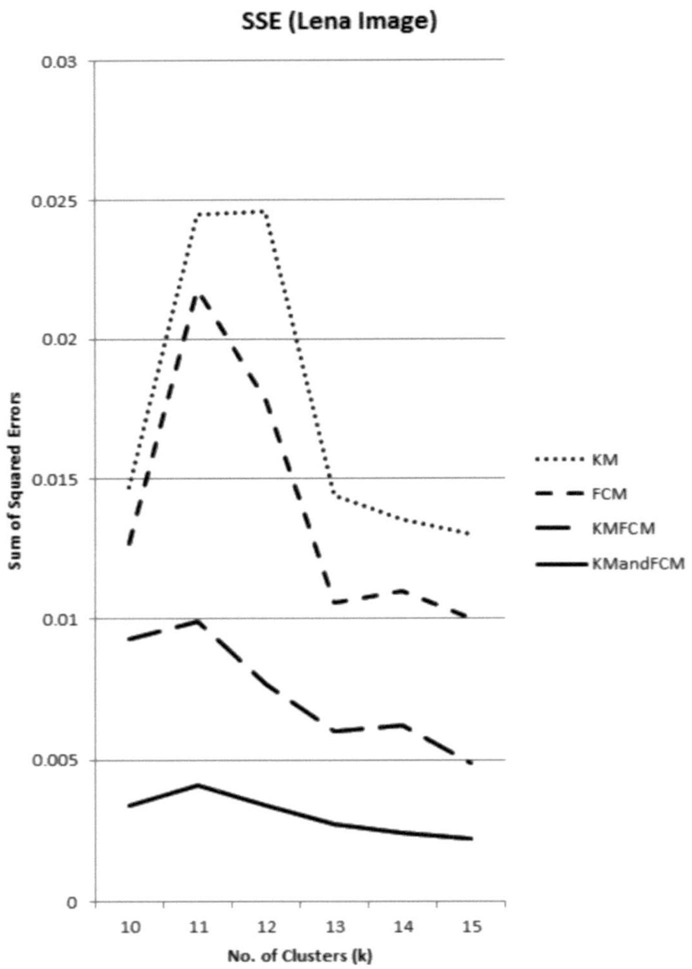

Figure 12.
Sum of squared errors (Lena image).

6.5 Original images used for present experimentation

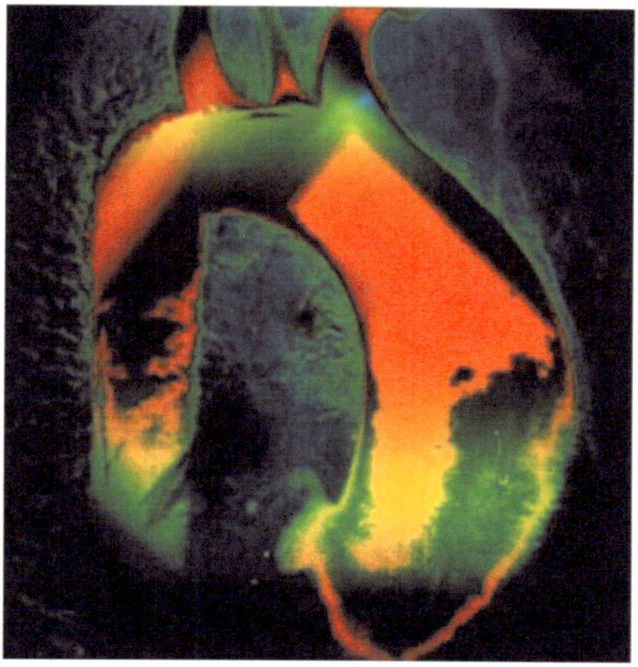

Figure 13.
Heart image.

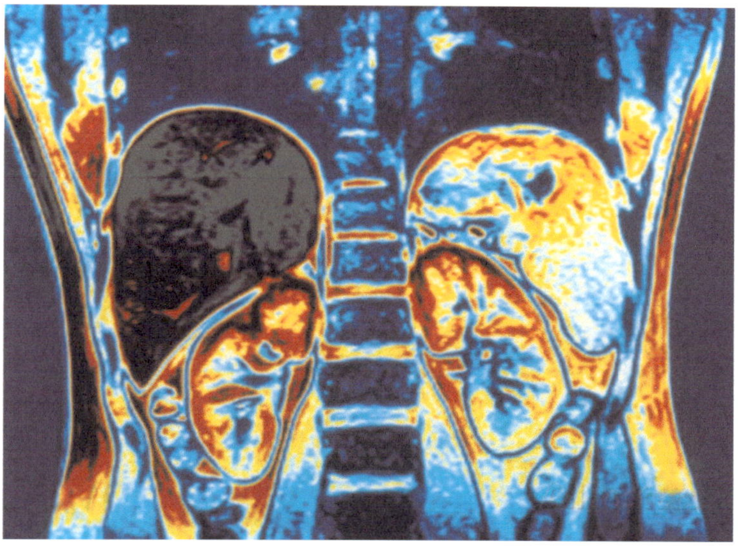

Figure 14.
Kidneys image.

Figure 15.
Baboon image.

Figure 16.
Lena image.

6.6 Comparison of segmentation results on Baboon image

As an example of the present experiments for image segmentation, segmentation results for Baboon image for 10 clusters are presented here. These results are generated by the above proposed hybrid clustering algorithms along with the standard K-Means and standard *FCM* algorithms.

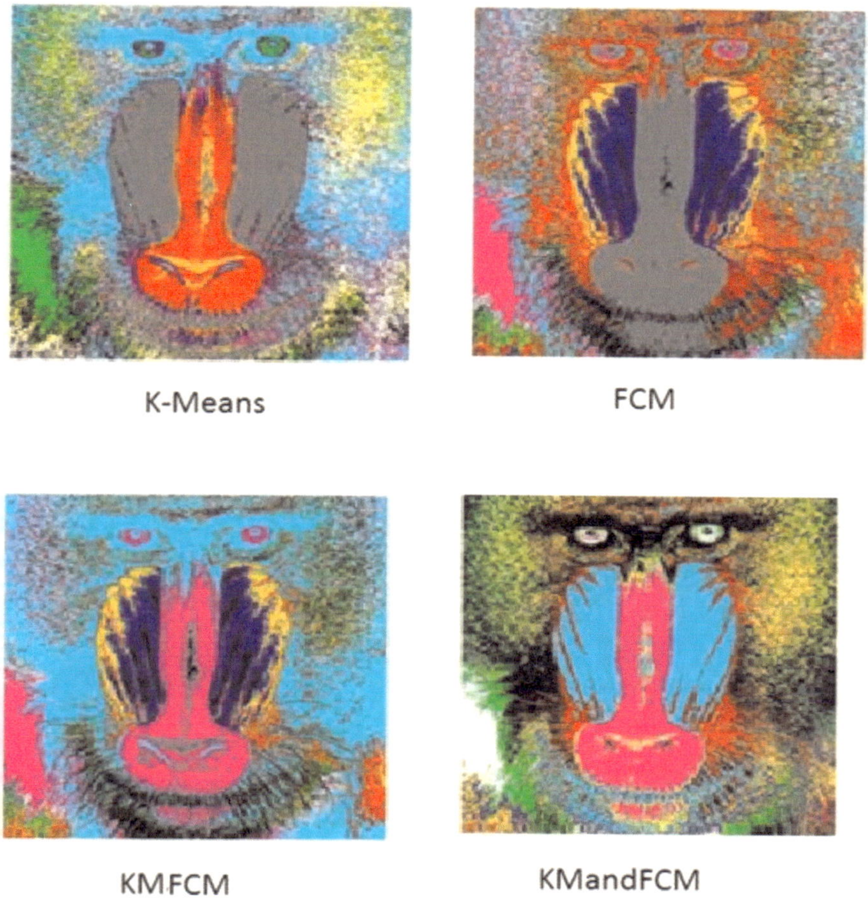

Figure 17.
Image segmentation results for Baboon image (for 10 Clusters).

For segmentation, here, each algorithm is executed using Baboon image data assuming that the number of clusters is 10, i.e., $k = 10$. Each segment is represented by each cluster. Separate color code is assigned to each cluster. The color codes are red, yellow, green, blue, orange, black, white, gray, cyan and magenta. The projections of all segmentation results generated by the algorithms are shown in **Figure 17**. The original Baboon image also shown in the figure.

In all the experiments, it is observed that hybrid clustering algorithm *KMandFCM* is showing better performance in terms of CPU, clustering fitness and SSE than the other algorithms.

7. Conclusion

The present chapter notably includes the study of hybridization of popular clustering algorithms, K-Means and *FCM*, and identifies the best hybridization strategy. All experiments are carried out for segmenting four images, which include two medical images also. For all the algorithms CPU time, clustering fitness and sum of squared error (SSE) are taken into consideration while carrying out their performance evaluation. In all the experiments that are conducted, the proposed hybrid algorithm *KMandFCM* is exhibiting better performance in terms of CPU time, Clustering Fitness (CF) and SSE.

In all experiments, it is also observed that the proposed hybrid clustering algorithms are showing better performance than the standard K-Means and *FCM* algorithms. Especially the *KMandFCM* algorithm has good results when compared to all other algorithms. Thus, it could be concluded that the hybrid clustering algorithm *KMandFCM* will have good application in other fields too.

Author details

Raja Kishor Duggirala
Department of CSE, Dr. L. Bullayya College of Engineering (for Women), Visakhapatnam, A.P., India

*Address all correspondence to: rajakisgor@gmail.com

References

[1] Kaltri K, Mahjoub M. Image segmentation by gaussian mixture models and modified *FCM* algorithm. The International Arab Journal of Information Technology. 2014;**11**(1): 11-18

[2] Wang S, Geng Z, Zhang J, Chen Y, Wang J. A fuzzy C-means model based on the spatial structural information for brain MRI segmentation. International Journal of Signal Processing, Image Processing and Pattern Recognition. 2014;**7**(1):313-322

[3] Khan AM, Ravi S. Image segmentation methods: A comparative study. International Journal of Soft Computing and Engineering (IJSCE). 2013;**3**(4):84-92. ISSN: 2231-2307

[4] Fwu J-K, Djuric PM. EM algorithm for image segmentation initialized by a tree structure scheme. IEEE Transactions on Image Processing. 1997;**6**(2):349-352

[5] Bezdek JC, Hall LO, Clarke LP. Review of MR image segmentation techniques using pattern recognition. Medical Physics. 1993;**20**:1033-1048

[6] Pham DL, Xu CY, Prince JL. A survey of current methods in medical image segmentation. Annual Review of Biomedical Engineering. 2000;**2**:315-337

[7] Wells WM, LGrimson WE, Kikinis R, Arrdrige SR. Adative segmentation of MRI data. IEEE Transactions on Medical Imaging. 1996;**15**:429-442

[8] Fraley C, Raftery AE. Model-based clustering, discriminant analysis, and density estimation. Journal of the American Statistical Association. 2002; **97**(458):611, ABI/INFORM Global

[9] Hannah IH, Selva KS. Hybrid TRS-FA clustering approach for Web2.0 social tagging system. International Journal of Rough Sets and Data Analysis (IJRSDA). 2015;**2**(1):70-87. DOI: 10.4018/ijrsda.2015010105

[10] Nizar Banu PK, Andrews S. Performance analysis of hard and soft clustering approaches for gene expression data. International Journal of Rough Sets and Data Analysis (IJRSDA). 2015;**2**(1):58-69. DOI: 10.4018/ijrsda.2015010104

[11] Ayramo S, Karkkainen T. Introduction to partitioning-based clustering methods with a robust example (Tech. Rep. No. C. 1/2006). Agora, Finland: University of Jyvaskyla, Department of Mathematical Information Technology; 2006

[12] Han J, Kamber M. Data Mining Concepts and Techniques, 2/e. New Delhi, India: Elsevier Inc; 2007

[13] Tan P-N, Steinbach M, Kumar V. Introduction to Data Mining, 1/e. New Delhi, India: Pearson Education; 2007

[14] Zadeh LA. Fuzzy sets. Information and Control. 1965;**8**(3):338-353

[15] Lemiare J. Fuzzy insurance. Astin Bulletin. 1990;**20**(1):33-55

[16] Chen S, Zhang D. Robust image segmentation using *FCM* with spatial constraints based on new kernel-induced distance measure. IEEE Transactions on Systems, Man, and Cybernetics. 1998;**34**:1907-1916

[17] Tsai A, Zhang J, Willsky AS. Expectation-maximization algorithms for image processing using multiscale models and mean-field theory, with applications to laser radar range profiling and segmentation. Optical Engineering. 2001;**40**(7):1287-1301

[18] Yang Y. Image segmentation by fuzzy C-means clustering algorithm

with a novel penalty term. Computing and Informatics. 2007;**26**:17-31

[19] Tolias YA, Panas SM. On applying spatial constraints in fuzzy image clustering using a fuzzy rule-based system. IEEE Signal Processing Letters. 1998;**5**:245-247

[20] Noordam JC, Van Den Broek WHAM, Buydens LMC. Geometrically guided Fuzzy C-Means clustering for multivariate image segmentation. In: Proc. International Conference on Pattern Recognition. Vol. 1. 2000. pp. 462-465

[21] Ahmed MN, Yamany SM, Mohamed N, Farag AA, Moriarty T. A modified fuzzy C-means algorithm for Bias field estimation and segmentation of MRI data. IEEE Transactions on Medical Imaging. 2002;**21**:193-199

[22] Zhang DQ, Chen SC, Pan ZS, Tan KR. Kernel-based Fuzzy clustering incorporating spatial constraints for image segmentation. In: Proceedings of International Conference on Machine Learning and Cybernetics; Vol. 4. 2003. pp. 2189-2192

[23] Li X, Li L, Lu H, Chen D, Liang Z. Inhomogeneity correction for magnetic resonance images with Fuzzy C-Mean algorithm. In: Proc. SPIE Int. Soc. Opt. Eng. Vol. 5032. 2003. pp. 995-1005

[24] Pham DL, Prince JL. Adaptive fuzzy segmentation of magnetic resonance images. IEEE Transactions on Medical Imaging. 1999;**18**:737-752

[25] Wu X, Kumar V, Quinlan JR, Ghosh J, Yang Q, Motoda H, et al. Top 10 algorithms in data mining. Knowledge and Information Systems. 2007;**14**:1-37 (survey paper: DIO 10.1007/s10115-007-0114-2)

[26] Aggarwal N, Aggarwal K. A mid-point based k-mean clustering algorithm for data mining. International Journal on Computer Science and Engineering. 2012;**4**(6):1174-1180

[27] Das S. Pattern recognition using fuzzy c-means technique, international journal of energy. Information and Communications. 2013;**4**(1):1-14

[28] Lu Y, Ma T, Yin C, Xie X, Tian W, Zhong SM. Implementation of the fuzzy C-means clustering algorithm in meteorological data. International Journal of Database Theory and Application. 2013;**6**(6):1-18

[29] Bezdek JC, Ehrlich R, Full W. *FCM*: The fuzzy c-means clustering algorithm. Computers & Geosciences. 1984;**10** (2–3):191-203

[30] Klir GJ, Yuan B. Fuzzy Sets and Fuzzy Logic: Theory and Applications. Upper Saddle River, New Jersey: Prentice Hall of India Private Limited; 2005

[31] Bezdek JC, Ehrlich R, Full W. FCM: The Fuzzy c-Means Clustering Algorithm. Computers & Geosciences. 1984;**10**(2-3):191-203

[32] Bezdek JC. Feature selection for binary data-Medical diagnosis with fuzzy sets. In: Proc. Nat. Comput. Conf. AFIPS Press; 1972. pp. 1057-1068

[33] Cannon RL, Dave JV, Bezdek JC. Efficient implementation of fuzzy c-means clustering algorithm. IEEE Transactions on Pattern Analysis and Machine Intelligence. 1986;**8**(2):248-255

[34] Cannon RL, Jacobs C. Multispectral pixel classification with fuzzy objective functions. Cen. for Automation Res., Univ. Maryland, College Park, Tech. Rep. CAR-TR-51. 1984

[35] Gong M, Liang Y, Ma W, Ma J. Fuzzy C-means clustering with local information and kernel metric for image segmentation. 2013;**22**(2):573-584

[36] Krinidis S, Krinidis M, Chatzis V. Fast and robust fuzzy active contours. IEEE Transactions on Image Processing. 2010;**19**(5):1328-1337

[37] Yong Y, Chongxun Z, Pan L. A novel fuzzy C-means clustering algorithm for image thresholding. Measurement Science Review. 2004; **4**(1):11-19

[38] Hanesch M, Scholger R, Dekkers MJ. The application of fuzzy C-means cluster analysis and non-linear mapping to a soil data set for the detection of polluted sites. Physics and Chemistry of the Earth (Part A). 2001;**26**(11–12): 885-891. DOI: S1464-1895(01)00137-5

[39] Adebisi AA, Olusayo OE, Olatunde OS. An exploratory study of K-Means and expectation maximization algorithms. British Journal of Mathematics and Computer Science. 2012;**2**(2):62-71

[40] Kearns M, Mansour Y, Andrew YNg. An Information-theoretic analysis of hard and soft assignment methods for clustering. In: Uncertainty in Artificial Intelligence, 1997. pp. 282-293

[41] Ghosh S, Dubey SK. Comparative analysis of K-means and Fuzzy c-means algorithms. International Journal of Advanced Computer Science and Applications. 2013;**4**(4):35-39

[42] Raja Kishor D, Venkateswarlu NB. A Behavioral study of some widely employed partitional and model-based clustering algorithms and their hybridizations. In: Proceedings of the International Conference on Data Engineering and Communication Technology, Advances in Intelligent Systems and Computing 469; Springer Nature; 2017. DOI 10.1007/978-981-10-1678-3_56

[43] Han X, Zhao T. Auto-K dynamic clustering algorithm. Journal of Animal and Veterinary Advances. 2005;**4**(5): 535-539

[44] http://www.biologyreference.com/Bl-Ce/Cardiovascular-Diseases.html

[45] http://www.allposters.com/-sp/Colour-MRI-Scan-of-Abdomen-Showing-Kidneys-Liver-Posters_i10024816_.htm?UPI=AP10024816_PC14258389_FI0_SV6_IT1_VRV1

[46] http://www.imageprocessingplace.com/root_files_V3/image_databases.htm

Chapter 9

The Software to the Soft Target Assessment

Lucia Mrazkova Duricova, Martin Hromada and Jan Mrazek

Abstract

The soft targets are closely related to the risk of attack to the group of people (to the lives). This problem can cause fatal consequences for the population. The current situation on the world reflects the fear of the attack in the soft targets. We can see the fear to lose life at these public places and in all types of access to free buildings. Each of us spends time in the shopping centers or the park every day, and our children spend time in schools where they can be threatened. The characteristics between the soft targets belong to a considerable number of persons at the same time in the same area, and the current state of the security measures is not adequate to the threats yet. The main aim of the software to the assessment of the soft target is to protect the people in the soft targets, minimize the impact to the people (visitors), and help to solve the problem at the moment. The methodology is based on the assessment of the object according to the features (according to the criteria).

Keywords: soft target assessment, analysis, security coefficient, object analysis

1. Introduction

The soft targets, the crowded places, and the objects of critical infrastructure are very vulnerable to the risk of the attack. These objects are identified as objects with a large number of visitors per day, and the situation is not as secure as the object requires [1]. These visitors are the aim of the potential attacker. The potential attacker is a man, who believes that the attack can solve the problem (problem with the political situation, the problem with the state system, the problem with the world). The attacker believes that the fear of death can solve the problem or can spread fear between other people. The attack on soft targets can disrupt the functionality of the state with fatal consequences [2].

The proposed methodology and the software tool can help us to the evaluation process of the soft targets and can help us to protect the life of the people. The main problem is that these places have open access to people all day. The open access can cause that the early detection of the attacker is difficult and a lot of people are at risk. If we can analyze the features of the object (in advance), then we will solve the problem effectively [3].

The current state of the research is described in this chapter. We have developed the static part of the assessment tool, and this static part was verified in the practice use (analyses of the soft targets in the Czech Republic). In Section 2, the attacks in the last years are described. The mathematical definition of the analysis of the soft targets is described in Section 3. Section 4 describes the case study of the shopping centers in the Czech and Slovak Republic. The case study of the train and bus station is described in Section 5. Finally, we describe the final comparison between the shopping center analyses and the train and bus station analyses in Section 6. Finally, we summarized the conclusion in the last section.

2. The attacks in the last year

In **Figure 1**, you can see the timeline of the last attacks from the 2019. This timeline is focused on the attacks to the civilians. These attacks were in the soft targets and crowded places realized.

As you can see in **Figure 1**, attacks on the soft targets killed 139 persons at least, and 301 people were injured. We can say that attacks on soft targets are very popular. On the other hand, a lot of attacks on the soft targets were revealed before the attacker will fulfill the attack. We can say that our system is more needed in the last years.

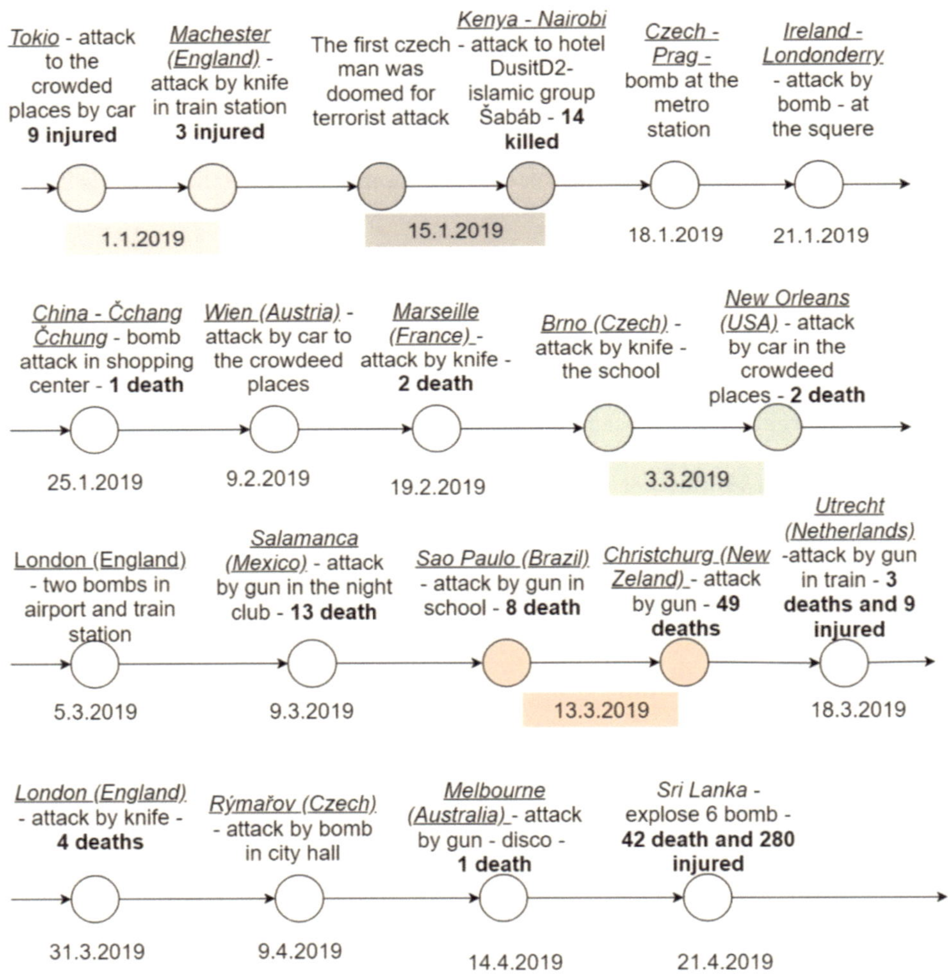

Figure 1.
The timeline of the attacks in 2019.

3. The mathematical definition of the analysis of the soft targets

The whole analysis of the features of the soft target is based on the relation between the question and the answer. The question analyzes the status of the features of the object. Each of the answers can define the level of the security, the level of the question, and level of the features too. This fact can be seen in **Figure 2**.

The level of the security can be influenced by the security measures. After the repeated assessment, we can see the higher level of the security.

The whole coefficient of the object is defined in the next equations:

$$K_S = \frac{L.W_1 + C_{EK}.W_2 + C_{PK}.W_3 + C_{IK}.W_3}{4} \quad (1)$$

K_S—the final security coefficient; Wn—weight of each coefficient; L—coefficient of the locality; C_{EK}—final coefficient of the exterior; C_{PK}—final coefficient of the processes; C_{IK}—final coefficient of the interior.

The weight (Wn) is set by the administrator of the software tool. We propose that these weights will be clarified after more case studies. In the current research, we evenly set these weights. The locality coefficient is defined by the map tool. The map tool has defined the risk of the locality by the administrator.

The coefficient of the interior is defined in Eq. (2). Each of the security attributes can be used in the object several times. The use of these security attributes is significant to the whole interior security. The security attributes define how many times the security attributes can be used:

$$I_K = \frac{1}{PB} \sum_{i=1}^{PB} B_i, \ (B_i \in < 0, 100) \quad (2)$$

PB—the number of the security attributes; B_i—the security attributes; I_K—the criteria of the interior.

Eq. (3) defines the final interior criteria in the same category of the interior. The final interior criteria (category) are based on the sum of each of these categories of the interior criteria:

$$I_{KCi} = \frac{1}{PK} \sum_{i=1}^{PK} I_K \quad (3)$$

I_{KCi}—interior criteria (all); PK—the number of the criteria.

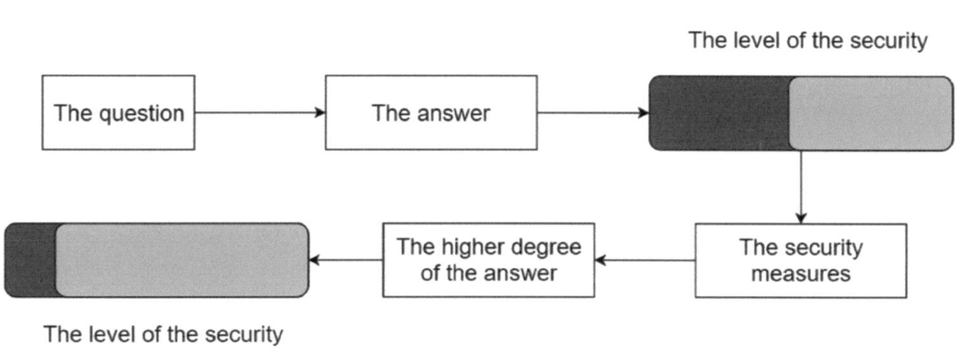

Figure 2.
The basics of the analysis of the soft targets.

The categories of the interior criteria are incorporated to the final coefficient of the interior. This equation is defined in Eq. (4):

$$C_{IK} = \frac{1}{P}\sum_{i=1}^{P} I_{KCi} \tag{4}$$

C_{IK}—the final coefficient of the interior; P—the number of up criteria.

The final coefficient of the interior is the sum of all the up criteria. The exterior coefficient is defined in Eq. (5):

$$E_K = \sum_{k_u=0}^{3} N * k_u = n * 0 + n * 1 + n * 2 + n * 3 \tag{5}$$

E_K—the calculation of the value of the exterior criteria; k_u—coefficient of the security level; N—the number of the attributes.

$$E_{KC} = \frac{\sum_{ku=0}^{3} n * k_u}{3 * \text{sum N}} \tag{6}$$

E_{KC}—all of the criteria of the exterior; Sum N—the number of all attributes.

$$\text{sum N} = \sum_{i=1}^{4} N_i \tag{7}$$

The final coefficient of the exterior is defined in Eq. (8).

$$C_{EK} = \left[\frac{10}{n}\sum_{j}^{n} E_{KCj}\right] * W \tag{8}$$

C_{EK}—the whole coefficient of the exterior.

Each of these equations is used in the next part of the paper (in the case study). The process coefficient is defined in Eq. (9):

$$P_K = \sum_{k_u=1}^{3} n_k * k_u \tag{9}$$

P_K—coefficient of one process (the number of criteria); nk—the number of the criteria; k_u—the level of the criteria.

$$P_{KCj} = \frac{10}{3 * N}\sum_{i=1}^{n} P_{Ki} \tag{10}$$

P_{KCj}—the complete process coefficient of the all processes in one category (processes are divided into the categories); N—the number of the processes; n—the number of the upper level of the process; P_{Ki}—each of the coefficient of one process.

The complete process coefficient is defined in Eq. (11). Each category has defined the weight according to the threats, or we can evenly set the weight:

$$C_{PK} = \left[\frac{1}{N}\sum_{j=1}^{k} P_{KCj}\right] * W \tag{11}$$

CPK—the whole coefficient of the processes; k—the criteria; W—the weight of the process.

This part of the chapter defined the mathematical definitions of the whole process of the analysis. We have defined three types of concrete analysis (processes, internal, and external) and one type of outside analysis (locality coefficient). The locality coefficient is defined according to the situation in the nearest area of the object. We can say that the locality can be changed in time without the change in the object. For example, the public event can influence the security situation in the object (e.g., the Christmas market).

4. The case study of the analysis of the shopping center

The case study was done to 35 objects which are situated in the Czech Republic and Slovak Republic. The objects which were analyzed belong to the next types of the categories: shopping centers, schools and universities, authority offices, train and bus stations, multifunctional buildings, sports stadiums, hospitals, and theaters and cinemas. In this part of the chapter, we are talking about the analysis of the shopping centers.

In **Figure 3**, we can see the analysis of nine shopping centers. On the x-axis, the object number is defined. On the y-axis, the security coefficient is defined. Value 10 represents the best security situation in the object. On the other hand, value 0 or 1 represents the worst security situation in the object.

As you can see in **Figure 3**, the best security situation is object 2. The object has the highest final security coefficient (KS) with value 6.5 (CIK). The final coefficient of the interior has object 2 with value 8.54. This value is the best security interior coefficient from this case study too. Exterior and process coefficient is the best value from the case study in object 2 too. In the next part, we can see the characterization of object 2.

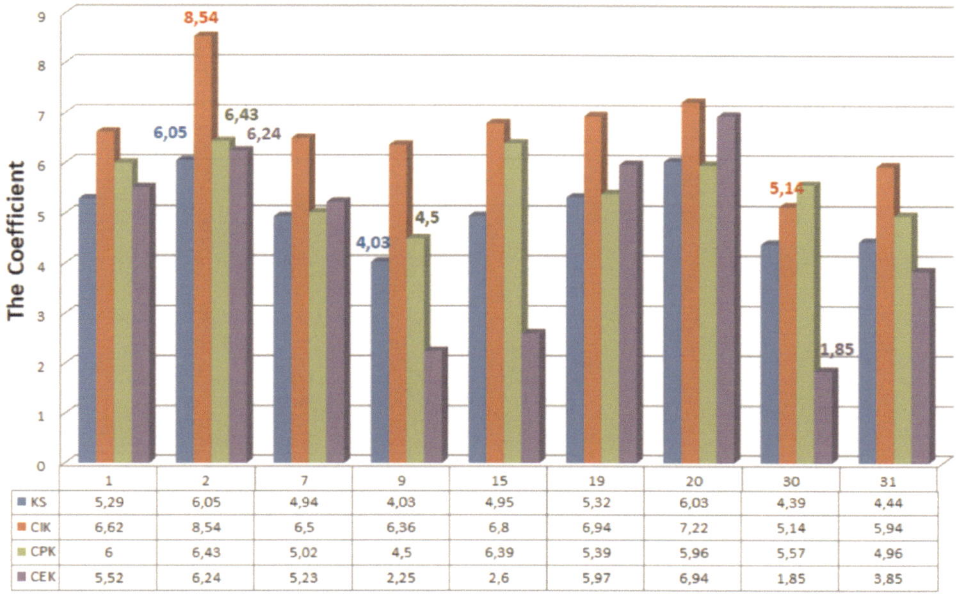

	1	2	7	9	15	19	20	30	31
KS	5,29	6,05	4,94	4,03	4,95	5,32	6,03	4,39	4,44
CIK	6,62	8,54	6,5	6,36	6,8	6,94	7,22	5,14	5,94
CPK	6	6,43	5,02	4,5	6,39	5,39	5,96	5,57	4,96
CEK	5,52	6,24	5,23	2,25	2,6	5,97	6,94	1,85	3,85

Figure 3.
The analysis of the shopping centers.

In **Figure 4**, we can see the locality of object 2. Object 2 is a relatively new shopping center. The locality is not in the center of the city, but the locality is very important for the visitors. Around the object is a river. The locality and the city are not so visited per day by visitors, and this fact can have a significant impact on the assessment of the locality and exterior. We need to say that this fact can be in the time variable. For example, the visit of the president can make the locality more unsafe, and the object will have a lower security coefficient.

In **Figure 5**, we can see antiterrorism pillars that are installed before entering into the object. Object 2 has integrated a lot of security measures because it is a relatively new object.

As you can see in **Figure 5**, these pillars can protect the enter to the object against the attack by car (drive the vehicle into the people).

On the other hand, the worst security situation is in object 9. This object can be seen in **Figure 6**. Object 9 is in the other city in the Czech Republic. This object is not in the center of the city, but this object is very close to the middle of the center. Object 9 is close to the main road, which is located across the parts of the city. Object 9 is located around the Rock café; in Rock café the rock concerts and the similar actions are organized too. This object can cause the risk of the violent attack to the object, however not to the people, because these actions are in the evening hours, when the shopping center is closed. This object has turnpike before its enter to the parking places.

As we can see in **Figure 3**, object 9 has a lower exterior coefficient. This fact can be caused by a close distance to the main road, the middle of the center, and the Rock café too.

This part of the chapter described the results of the case study in the shopping centers as a category. In the next part, we can describe the result of the train and bus

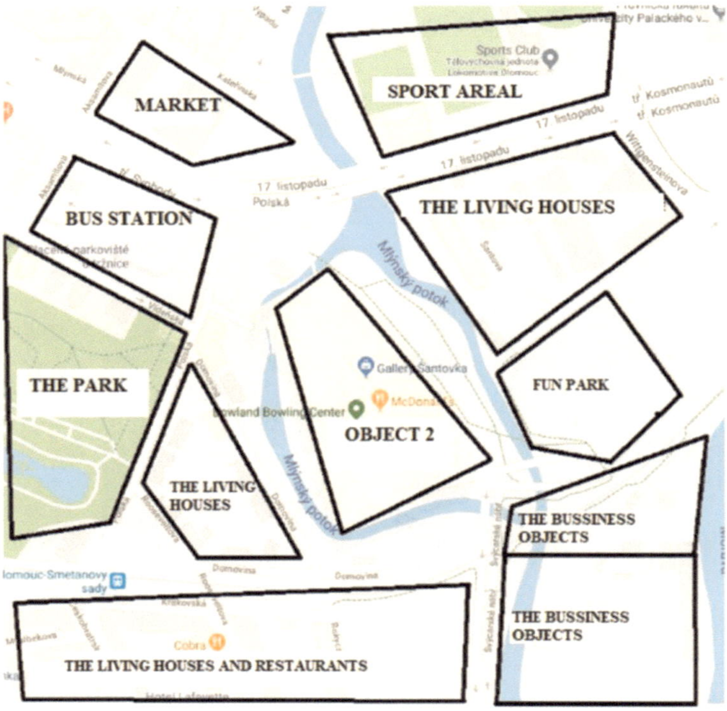

Figure 4.
The locality of object 2 (shopping center).

Figure 5.
The antiterrorism pillars before the enters.

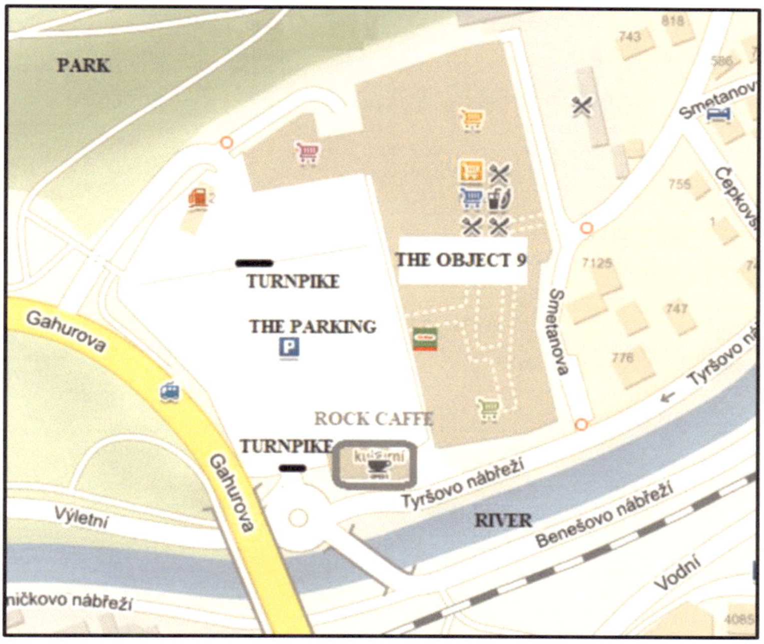

Figure 6.
The locality of object 9.

station. These objects are opened, and we can see the differences between two types of security coefficients which are caused by the different categories.

5. The case study of the analysis of the train and bus station

This chapter describes the analysis of the objects in the category train and bus station. We can expect that these objects will have the differences between the exterior coefficient and the shopping centers. If the case study confirms this theory, we can say that the proposed methodology to the analysis corresponds to the reality.

In **Figure 7** we can see the objects (train and bus stations) which are analyzed in this case study. The numbers in **Figure 6** mark the number of objects.

In **Figure 8**, we can see the data of the case study. The best security situation is object 14. This train and bus station is located in Frydek-Místek. This result is caused by the analysis. The analysis was realized only in the building of the bus station. The analytics did not analyze the train station, which shows that the final coefficient will be lower. On the other hand, the worst situation according to the final security coefficient is object 3. Object 3 is located in Zlin. This city is a country town. We can say that this object was analyzed correctly.

Object 3 is oriented in the middle of the center. The rail tracks are going across the middle of the town. On the other hand, we need to say that rail transport is not so used in Zlin. In **Figure 9**, we can see the localization of the train and bus station.

Finally, we can say that this analysis in the category train and bus station has some differences between assessments. This fact is caused by the aim of the analysis and software tool. The proposed software tool was developed for analyzing the buildings and not for analyzing open spaces. The second reason can be that the case study was done with more analytics.

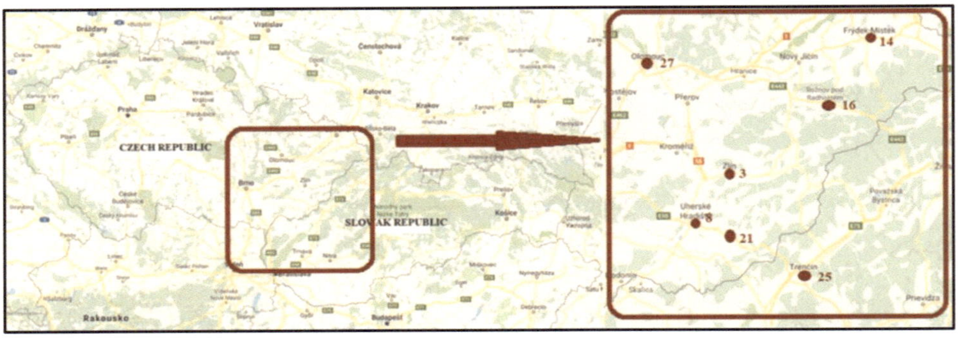

Figure 7.
The localization of the train and bus stations.

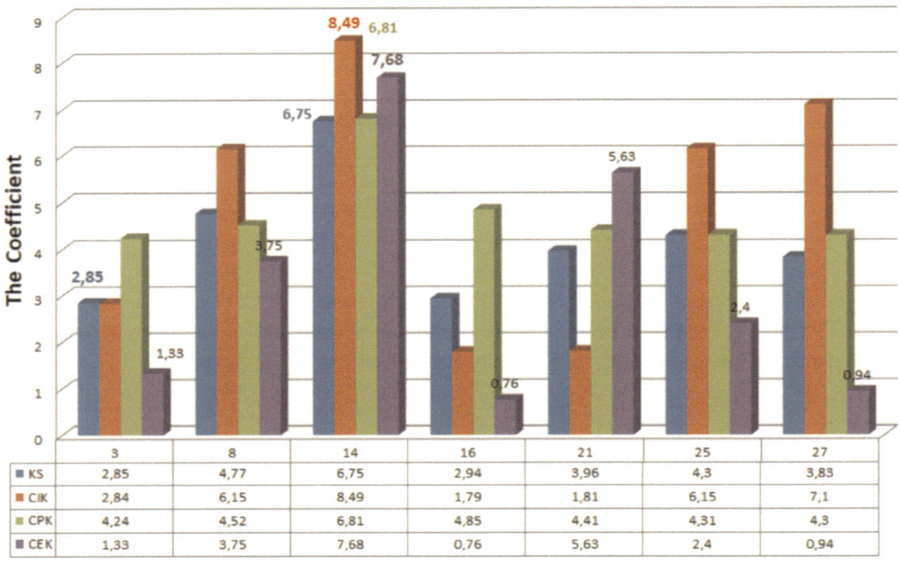

	3	8	14	16	21	25	27
KS	2,85	4,77	6,75	2,94	3,96	4,3	3,83
CIK	2,84	6,15	8,49	1,79	1,81	6,15	7,1
CPK	4,24	4,52	6,81	4,85	4,41	4,31	4,3
CEK	1,33	3,75	7,68	0,76	5,63	2,4	0,94

Figure 8.
The analysis of the train and bus station.

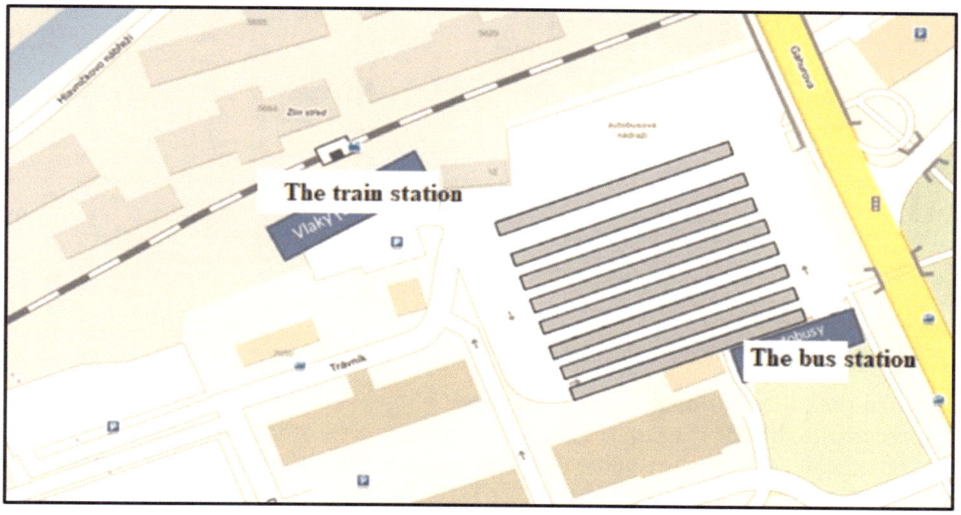

Figure 9.
The locality of the train and bus station in Zlin.

6. The final comparisons

We can constant that each of the objects (train and bus stations) has a different exterior security coefficient in contrast with the shopping centers. We can see this contrast in **Figure 9**.

As we can see in **Figure 10**, the difference between the average values is significant. The average value of the exterior coefficient in the shopping centers is 4.49. The average value of the exterior coefficient in the train and bus station is 1.75. As we can see on the right side of the figure, the train and bus station has significant differences in the two cases. Object 21 and object 14 are the objects that have not been analyzed correctly. The analyst analyzes only one part of the station. In these two cases, we do not use the average value. We can constant that each of the objects (train and bus stations) has the different exterior security coefficient in contrast with the shopping centers. We can see this contrast in **Figure 10**.

As we can see in the right side of **Figure 11**, object 14 (train and bus station in Frýdek-Místek) is not correctly analyzed. We cannot use these results as the correct

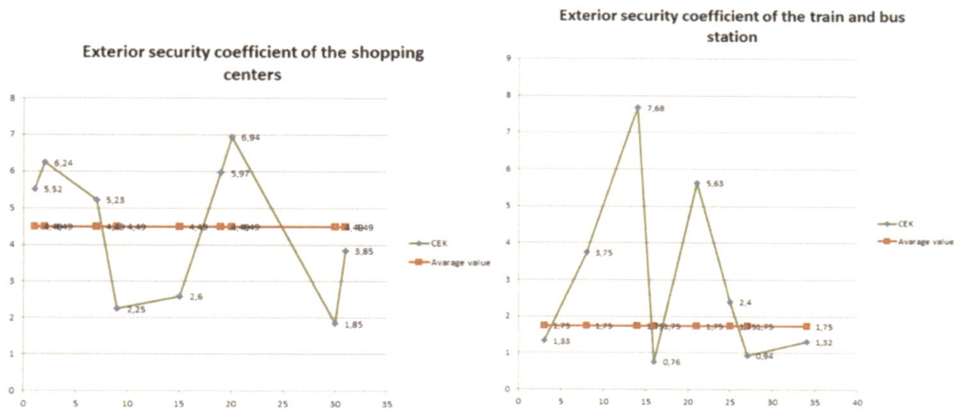

Figure 10.
The comparison between shopping center and bus and train station in exterior coefficient.

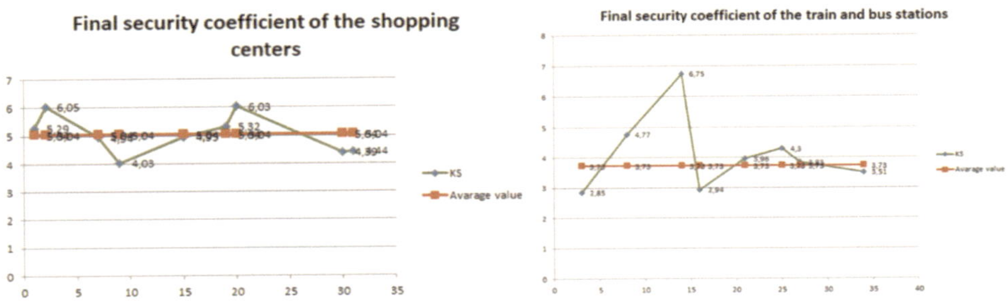

Figure 11.
The final comparison between the train and bus stations and shopping centers.

data. On the other hand, object 8 is the second best security coefficient. This object is correctly analyzed. We can see that the average value of this analysis is 3.73. On the other hand, the average value of the analysis of the shopping center is 5.04. We can say that the analysis corresponds to the reality because open-space objects will have a significant difference in security situations.

7. Conclusion

This chapter aimed to describe the proposed software tool which is realized on the website www.softtargets.eu. This methodology was developed as a doctoral thesis. In the first part of this chapter, the mathematical definitions of the proposed methodology were described. The second part of this chapter, the case study, was described. This case study was oriented to the comparison between shopping center analysis and the train and bus station analysis. The results of the shopping centers and the results of the train and bus station have significant differences. These differences were caused by exterior differences: the shopping center analyses (the building and close places) and the train and bus station analyses (the open spaces). We can constant that the proposed software can be used to the analysis of the open spaces too. However, we need to analyze all buildings and all aspects of the open spaces. We need to set more criteria for the open spaces. These criteria will be different from the criteria of the buildings because these criteria have to study the other features of the soft target.

Acknowledgements

This chapter is realized as the research with a doctoral student. This work was supported by the Internal Grant Agency of Tomas Bata University in Zlin under project no. IGA/FAI/2019/009; by the research project VI20172019073 "Identification and methods of protection of Czech soft targets against violent acts with elaboration of a warning system," supported by the Ministry of the Interior of the Czech Republic in the years 2017–2019; and by the research project VI20172019054 "An analytical software module for the real-time resilience evaluation from point of the converged security," supported by the Ministry of the Interior of the Czech Republic in the years 2017–2019.

Author details

Lucia Mrazkova Duricova*, Martin Hromada and Jan Mrazek
Faculty of Applied Informatics, Tomas Bata University in Zlin, Zlin, Czech Republic

*Address all correspondence to: lmrazkova@utb.cz

References

[1] Hesterman JL. Soft Target Hardening: Protecting People from Attack. Vol. xxii. Boca Raton: CRC Press, Taylor & Francis Group; 2015. p. 299. ISBN 978-1-4822-4421-2

[2] Rosenberg F. Nice Solutions for Critical Facilities. Nidam: NICE; 2014

[3] Fekete A. Safety and security target levels: Opportunities and challenges for risk management and risk communication. International Journal of Disaster Risk Reduction. 2012;**2**:67-76

Chapter 10

The Methodological Standard to the Assessment of the Traffic Simulation in Real Time

Jan Mrazek, Martin Hromada and Lucia Duricova Mrazkova

Abstract

The quantity of goods transported in the transport sector is increasing every year. As a result of the increase, the number of means of transport increases. The most popular sector is road transport, which is also referred to as the most dangerous in terms of safety. The assessment of the traffic situation on the planned route does not take place during its implementation. The consequences of long reaction times on emerging or already occurring incidents affect safety. This phenomenon can also trigger crisis situations in other critical infrastructure sectors. In more serious events, a cascading effect can occur between critical infrastructure elements that could lead to a domino effect. This phenomenon could be likened, for example, to *blackout* in power engineering. The conclusion of the chapter will include a case scenario as to how a methodological standard for traffic assessment should work on real-time crises.

Keywords: critical infrastructure, transport, road transport, crisis situations, traffic, evaluation criteria, traffic modeling

1. Introduction

The chapter is focused on the issue of crisis situations in the transport of material in road transport. The most important elements in transporting transport units are time and money at the moment. Time is taken from the point of view of when and where to pick up the shipments. Of course, time-related information is important where and when to ship the shipment to the destination. Based on this information, a preliminary cost calculation can be performed. The financial valuation depends not only on the distance to which the transport is carried out. Another important information is whether it is a special transport unit and whether special measures or training are needed to transport it (e.g., transport of dangerous substances—transport of ADR in road transport), etc. transport units to quantify the real costs of transporting a transport unit from point A to point B [18, 19].

Thousands of tons of material are shipped around the world every day. The first part of the book is focused on the development of transport units in individual transport sectors in the Czech Republic. In this chapter, we will learn not only the most popular transport sector for transporting transport units. Based on developments in the years, it will be possible to prepare preventive measures to minimize risks. Risks in transport are understood to be situations disrupting the transport process. Risks can be characterized as influenced or unaffected. The proposed tool should work with both to be an effective tool for both planning and solutions in the ongoing transport process [11–13, 15].

In the conclusion of the thesis, a case study is described. The case study focuses on preventive measures before the emergence of a crisis situation. The proposed tool should be capable of timely response to the tool operator. Based on the information obtained, the tool suggests options for solving crisis situations before and after its creation. The proposed tool works with input data throughout the process. The aim of the instrument should be to prevent but also respond to crisis situations [11, 12].

2. Transport unit in transport

Transport units in transport can be divided into two parts. The first part is focused on the transport of people, which is not so important for the solution of crisis situations. The number of passengers carried varies across countries. Priority for public transport is preferred for passenger transport in developed countries. This situation in the Czech Republic is rather the opposite, and passenger traffic wins over the public. The trend transferred from abroad in the form of limiting the parking of cars in the centers, and their close surroundings began to be introduced in the Czech Republic. These limitations come under the name of a parking zone in cities. With the help of parking zones, however, the problem is gradually delayed, not the solution. This idea does not lead to the solution of the problem and can result in overcrowding of the periphery of the city [16].

2.1 Transport of people

In **Table 1**, we captured the number of passengers carried in years in each sector. The table shows the development in individual sectors and at the same time presents their popularity. Water transport in the Czech Republic has no representation in passenger transport. For this reason, no measurable data is recorded when transporting people.

The table shows us the downward number of passengers. We will present this fact in **Figure 1**, which presents a downward trend since 2015.

In **Figure 1**, we can observe a downward trend that started in 2015 and will probably not be different in 2018 and 2019. This fact shows us the decreasing popularity of public transport in passenger transport. This phenomenon is mirrored in popularity and preference by our own means of transport. At the same time, this figure highlights the increasing number of cars on the roads. As a result of the increasing number of cars on the road, there is also a higher density and thus an increased risk of crisis.

Passenger transportation (in thousand)	2014	2015	2016	2017
Road transport	349,515	350,920	332,763	329,733
Rail transport	176,051	176,624	179,171	183,024
Air transport	5623	5393	6000	6657
Water transport	—	—	—	—

Table 1.
Passenger transportation in the Czech Republic [1].

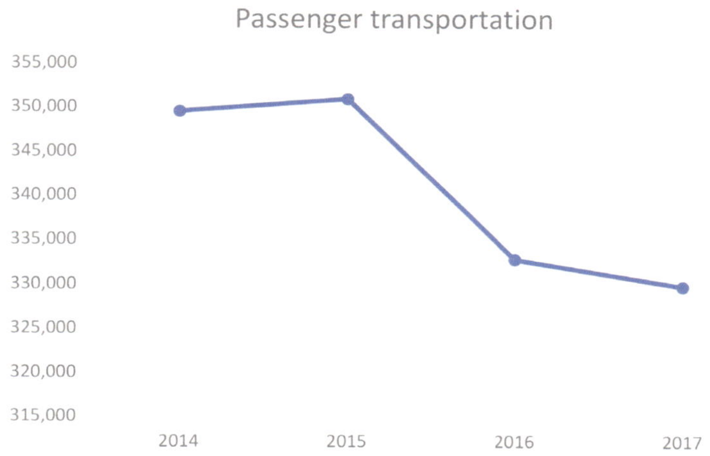

Figure 1.
Graph of development of transported persons in the Czech Republic [1].

Thinks (in thousand)	2014	2015	2016	2017
Road transport	386,243	438,906	431,889	459,433
Rail transport	91,564	97,280	98,034	96,516
Air transport	9057	5790	5632	6362
Water transport	1780	1853	1779	1568

Table 2.
Amount of transported material in the Czech Republic [1].

The chapter in the book is mainly devoted to the second part. The transport of material is more popular every year, and this is confirmed by the fact how much material is transported every year.

2.2 Transport of material

In **Table 2**, we can see statistics from the transport sectors. These data capture the number of transported transport units in each transport sector. Water transport is also one of the transport sectors. This sector is used for transporting material but not for transporting people due to watercourses on which people could be transported [1, 2].

Table 2 shows the dominance of road transport in material transport. Road traffic is the only increase. Other sectors are stagnating. This phenomenon can be seen in **Figure 2**.

Material transport is specific not only to the customer but also to the carrier. The growth in the transport sector is due to the fact that as the only transport sector, material from point A to point B can be transferred. Other sectors except the road sector have transport from terminal to terminal.

In conclusion, when observing statistical data, it can be stated that road transport is the busiest transport sector. At the same time, heavy road traffic increases the risk due to the number of vehicles traveling. The increasing number of means of transport during the most popular times increases the possibility of traffic accidents, which considerably complicate the process of material transport [3, 4].

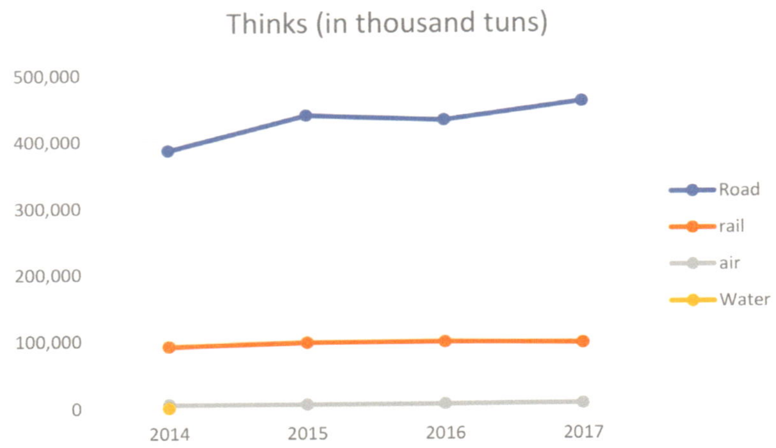

Figure 2.
Development of transported goods in the Czech Republic in individual transport sectors [1].

3. Transportation

The material is transported in each transport sector. As presented in point 2.2, road transport in the amount of transported material is clearly determined. From the point of view of the most popular form used to transport material, our tool is also focusing on this sector. The risks associated with the transport of material are largely predominantly in terms of responding to a new crisis situation [5].

The shortcomings of the present are seen not only in terms of transport planning but also during the entire transport process. The planning process does not work with many dynamic data that can greatly affect the process. These data include, for example, weather development, traffic density on a planned route, road reconstruction, risks associated with traffic accidents, etc. [14].

3.1 Current status

The current state of the planning phase does not work with a great deal of substantial information present. Common shipping cost calculations are:

- Distance
- Amount
- Price

With the above data, market demand is created. This demand has, in the final analysis, a predominantly price criterion. The cheaper the more interesting for customer.

3.2 Application of the proposed tool

Input data is important for each tool or software. He should work with a lot of data for our material transport management tool. This process makes it a dynamic tool that can work and respond in real time to crisis situations. Crisis situations can emerge until extraordinary events, which in turn cripple other elements of critical infrastructure [6, 7] (**Table 3**).

Type of transport unit	• Valuables • Fragile • People • Oversized cargo • Dangerous cargo • Others • Food • Prisoners • Combination
Standardization of the type of transport unit	• Yes • No
Disposition of the transport unit	• Solid • Liquid • Gaseous
Dimension of goods	• Height • Width • Length • Weight • Quantity
Transport packaging	• Palette • Wooden barrel • Canister • Box • Bag • Fine metal bag • Combination of packaging
Packaging characteristics and restrictions	• The ability to pile packages on each other • Standardized package size • Number of packages
Categorization of the transport environment	• Dry environment • No limits
Choice of means of transport	• Motorcycle • Car • Truck
Other specifications and numeric operations	• Transport from A to B • Transport performance • Utilization of driving • Maximum carrying capacity of the vehicle • The size of the cargo area • Loading time • Vehicle emission class

Table 3.
Software input data [2, 8].

Setting the correct input data to be untidy for the setup is an important element. Its setting is narrative from the point of view of discussions with practitioners. Entering large amounts of input data discourages operators and dispatchers from using the tool. By defining only the most important input data, it will be possible to prevent the dangers of diversion of hazardous materials to cities, so as not to endanger the health and life of the population [9, 10].

Setting up input data can also lead to a reduction in the number of lorries or vehicles when transport units are combined. In the case of a combination of transport units that would be acceptable, two costs can be achieved by completing one transport which satisfies two requests. This model has been operating in America for many years. When merging transport units, you can not only reduce the number of cars but also reduce the CO2 share. This merging process has many other advantages, but we can also find disadvantages that are not as discouraging as possible. The more common input data can be avoided by jeopardizing the diversion of hazardous materials into cities to avoid threats to the health and life of the population [9, 10, 17].

Next, we will focus on a case study that will consist of the functionality of the proposed tool. The case study will describe how the proposed tool should function at each stage of the shipment. The description of each of them will outline what it should be most focused on at the same time to increase security and not create additional risks.

4. Case study

The case study focuses on the road transport sector. Because of the annual rise shown in **Figure 2**, this is the most important sector to minimize risk. At the same time, as the number of transported goods increases, there is an increase in risks as the number of means of transport also increases. Means of transport are a tool for the realization and fulfillment of customer requirements. Customer requirements are essential data for the proposed tool. Meeting the requirements is the most important point before starting the entire transport process.

4.1 Planning phase

In **Figure 3** we can see two points. Point A indicates where the transport unit is to be picked up. Point B shows us where it is necessary to transport the transport unit.

After entering the entry point to load the shipping unit and then determine the delivery location, the route selection will come in line. This phenomenon is shown in **Figure 4**.

Figure 4 shows the route options from which we can choose the most suitable for us. The tool works as a regular map, giving you the shortest route choices, the fastest or the alternative. The tool does not work based on common data such as a common map or navigation. By setting the input data of the so-called customer requirements, there is a direct definition of the routes that can be implemented in the transport process.

Based on dynamic functionality, the tool marks not only risks but also various constraints on routes, which could limit or disrupt the entire transport process. Reconstructions are usually given in advance together with a possible surface change. All this data is important to the functionality of the tool in order to be able to respond to columns and possibly a large delay on a planned route.

Figure 3.
Planning phase.

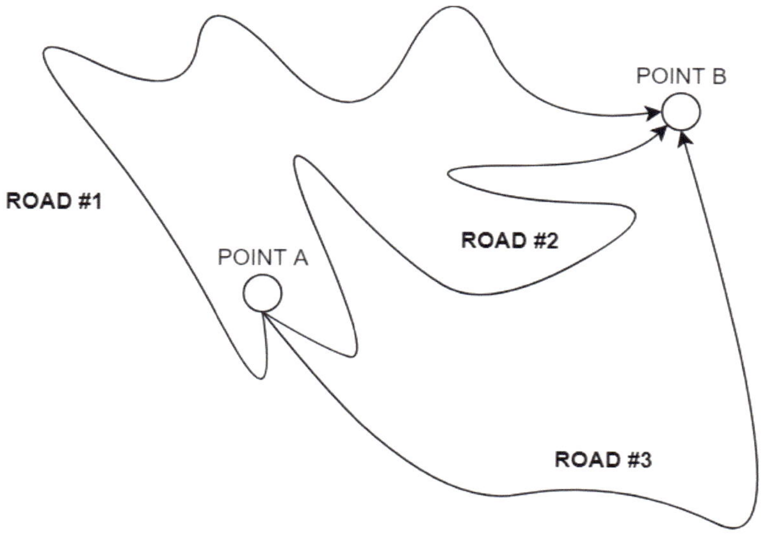

Figure 4.
Select the appropriate route.

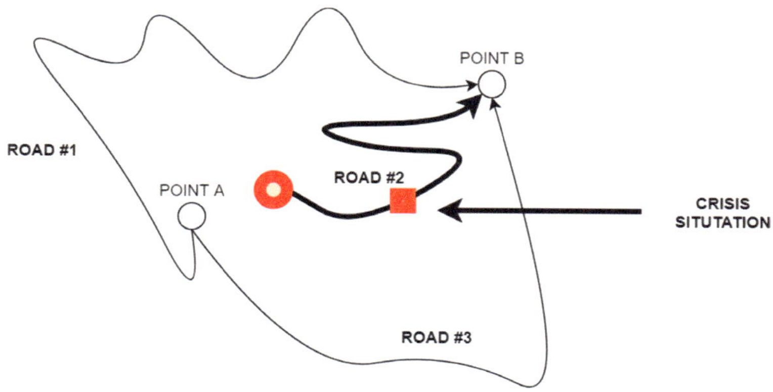

Figure 5.
Real phase during transport.

4.2 Real phase during transport

In this section, we will discuss the functionality of the proposed tool when working in real time. In **Figure 5**, we can see a situation that jumps to the operator during the transport process. The vehicle picked up the shipment or material and took the chosen route.

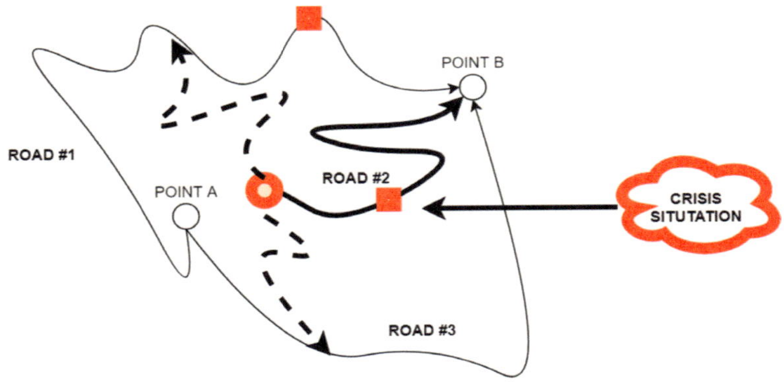

Figure 6.
Alternative solution to the situation.

During the transport process, a traffic accident occurred on the route. A traffic accident writes a delay of 4 hours in navigation. On the basis of the event, traffic is diverted to alternative routes using the police. Police divert traffic irrespective of material being transported in trucks. The categorization of cargo in this respect is not taken into account and thus endangers human health.

Immediately after the obstacle on the planned route, the operator or dispatcher using the software receives a problem report on the route. This phenomenon is shown in **Figure 6**.

After receiving a crisis situation, the operator receives a solution to the most appropriate solution to the traffic diversion. A diversion to alternative routes is to arrive at point B at an agreed time and to complete an alternative route so as not to pose a threat or other risks in the environment where the means of transport moves with the transport unit. All alternative solutions will alert us to the risks associated with the settings that were set at the input. In case we have marked the height of the means of transport by 3 meters on the routes, this will indicate I risk places in the form of a very small difference in the form of our possibility of limiting or creating another obstacle.

As the input data confirms, they can specify the entire transport process to create the most realistic and acceptable environment possible.

5. Conclusion

Material transport is important for every country. The interconnection of critical infrastructure elements points to prevention in each sector intertwined with another element. Prevention is an essential thing for a quick solution to an emerging or already occurring event in order to minimize the consequences. In case of minimizing the consequences, it is also necessary to think about the recovery time.

There is a wide range of risks that can be influenced or negligible. Impossible risks can include risky places from the point of view of traffic accidents, restrictions on a planned route, etc. Natural phenomena such as seasons, weather, and more are unimaginable.

It is not possible to predict possible crisis situations in the form of traffic accidents, traffic density, or weather. In planning, a route can be planned with the option of alternative, but dynamic driving during transport will deliver transport efficiency.

The dynamic part of the proposed tool works with input data but also with real data. Real data works at a time when the entire transport process takes place. Dynamic steering starts from the actual loading of the transport units to the means of transport.

The proposed instrument is still under development. All substantiated claims are supported by experts in this field. The number of transports increases every year, and this trend needs to be responded to avoid oversaturation of roads or other transport sectors. Road transport is the most popular, but, at the same time, it can be described as the most dangerous.

Another objective of the proposed tool should be to work with a large amount of data in the form of clustering based on traffic density, depending on their speed in the monitored area.

The risk of the functionality of the proposed tools is predominantly in its load. Working with large amounts of data could slow down the proposed tool considerably. Another risk in the form of use is seen in the form of input data; if there is a large amount of input data, it will potentially discourage potential users from using it.

Acknowledgements

This project is realized as the research with doctoral student, and it is the basic input for the next research, which we will develop in the next term. It was realized with the support of a university. This work was supported by the Internal Grant Agency of Tomas Bata University under the project no. IGA/FAI/2019/010.

This work was supported by the research project VI20152019049 "RESILIENCE 2015: Dynamic Resilience Evaluation of Interrelated Critical Infrastructure Subsystems" and by the Ministry of the Interior of the Czech Republic in the years 2015–2019.

Author details

Jan Mrazek*, Martin Hromada and Lucia Duricova Mrazkova
Tomas Bata University in Zlin, Zlin, Czech Republic

*Address all correspondence to: jmrazek@utb.cz

References

[1] Czech Statistical Office. Statistical Yearbook of the Czech Republic 2018. Czech Statistical Office 2018. Available from: https://www.czso.cz [cited: 2019-03-22]

[2] Mrazek J, Duricova L, Hromada M. The proposal of evaluation criteria for recoverability of road transport. In: Cepin M, Bris R, editors. Safety and Reliability—Theory and Applications. London: Taylor & Francis Group; 2017. ISBN: 978-1-138-62937-0

[3] Liu G, Lyrintzis AS, Michalopoulos PG. Modelling of freeway merging and diverging flow dynamics. Applied Mathematical Modelling. 1996;**229**(1996):317-345

[4] Juan DDO, Willumsen Luis G. Modelling Transport. John Wiley & Sons; 2011

[5] Zhang Y. Method and system to dynamically collect statistics of traffic flows in a software-defined networking (sdn) system. U.S. Patent Application No 14/462,444. 2016

[6] Zhang L et al. Deep potential molecular dynamics: A scalable model with the accuracy of quantum mechanics. Physical Review Letters. 2018;**120**(14):143001

[7] Ehsani M, Ahmadi A, Fadai D. Modeling of vehicle fuel consumption and carbon dioxide emission in road transport. Renewable and Sustainable Energy Reviews. 2016;**53**:1638-1648

[8] Mrazek J, Vavra J, Hromada M. The evaluation criteria in the road transported with fuzzy logic support. Annals of DAAAM & Proceedings. 2018;**29**

[9] Jiang J et al. Modelling traffic flows and estimating road travel times in transportation network under dynamic disturbances. Transportation. 2019:1-30

[10] Bullard RD. Transportation matters: Stranded on the side of the road before and after disasters strike. In: Race, Place, and Environmental Justice After Hurricane Katrina. Routledge; 2018. pp. 85-108

[11] Black J. Urban Transport Planning: Theory and Practice. Routledge; 2018

[12] Soysal M, ÇIMEN M, DEMIR E. On the mathematical modeling of green one-to-one pickup and delivery problem with road segmentation. Journal of Cleaner Production. 2018;**174**:1664-1678

[13] Wang Y et al. Dynamic traffic assignment: A review of the methodological advances for environmentally sustainable road transportation applications. Transportation Research Part B: Methodological. 2018;**111**:370-394

[14] Pyatkova K et al. Flood impacts on road transportation using microscopic traffic modelling techniques. In: Simulating Urban Traffic Scenarios. Cham: Springer; 2019. pp. 115-126

[15] Godbole S et al. Dynamic loading on a prefabricated modular unit of a building during road transportation. Journal of Building Engineering. 2018;**18**:260-269

[16] Chiou S-W. A traffic-responsive signal control to enhance road network resilience with hazmat transportation in multiple periods. Reliability Engineering & System Safety. 2018;**175**:105-118

[17] Liao T-Y, Hu T-Y, Ko Y-N. A resilience optimization model for transportation networks under disasters. Natural Hazards. 2018;**93**(1):469-489

[18] Cao X et al. Urban wasteful transport and its estimation methods. Sustainability. 2018;**10**(12):4562

[19] Crainic TG, Perboli G, Rosano M. Simulation of intermodal freight transportation systems: A taxonomy. European Journal of Operational Research. 2018;**270**(2):401-418

Chapter 11

Augmented Post Systems: Syntax, Semantics, and Applications

Igor Sheremet

Abstract

Augmented Post systems (APS) are string-operating Prolog-like knowledge representation, affiliated with the "Set of Strings" Framework (SSF). APS descriptive and logical inference capabilities provide natural integration of Big Data with online analytic processing. This chapter is dedicated to strict formal definition of APS syntax, mathematical and operational semantics, and to its most valuable implementational issues, as well as to APS application to Big Data, Internet of Things, cyberphysical industry, and cybersecurity areas.

Keywords: "Set of Strings" framework, augmented Post systems, string data bases, data and knowledge engineering, Big Data, Internet of Things, cybersecurity

1. Introduction

The background of "set of strings" Framework (SSF) [1–4], developed for adequate modeling of Big Data, is the representation of database as an updated Set of Strings, whose structure is defined by the current metadatabase (MDB), being a set of context-free generation rules in the sense of N. Chomsky [5].

Augmented Post systems (APS), considered in this chapter, are result of deep integration in the one toolkit of two well-known formalisms—classical string-generating formal grammars and string-operating logical systems, proposed by E. Post [6].

By this, the main features of APS are:

1. Deep connectivity with "Set of Strings" representation of databases.

2. Deductive capabilities, similar to those, which are inherent to various deductive extensions of the relational model of data, as well as to Prolog and Datalog [7–12].

The second feature provides selection of not only data, presenting in string databases (SDB) in the explicit form, but also data, which may be constructed (derived) from the explicit data by means of logical inference.

This chapter is dedicated to the formal definition and substantial description of APS and their applications. The terminology and abbreviations are the same, as have been used when describing SSF.

Section 2 is dedicated to the definition of syntax and mathematical semantics of APS, and Section 3 to their operational semantics. Procedural connection to APS is described in Section 4, and flow-processing APS (FAPS), based on the aforementioned connection, are described in Section 5. Implementation issues and applications of APS family are considered in Section 6. The conclusion contains proposals on further development of SSF.

2. Syntax and mathematical semantics of the augmented Post systems

The **augmented Post system** $P = <S, D>$ is the composition of set S of the so-called augmented, or string (S-), productions, and set D of context-free generating rules, being metadatabase. Following the terminology of knowledge engineering, we shall call P also as *APS-represented knowledge base* (for short, APS KB).

S-production $\sigma = <q, d>$ is couple, the first component of which, named *body*, has the form of postproduction:

$$s_0 \leftarrow s_1, \ldots, s_m, \tag{1}$$

where $m \geq 0$ and $s_0, s_1, \ldots, s_m$ are terms.

Term s_0 is called *header*, while terms $s_1, \ldots, s_m$ are called *conditions*. The set of conditions is unordered, i.e., S-productions:

$$<s_0 \leftarrow s_{i_1}, \ldots, s_{i_m}, d>, \tag{2}$$

and

$$<s_0 \leftarrow s_{j_1}, \ldots, s_{j_m}, d>, \tag{3}$$

where $\{i_1, \ldots, i_m\} = \{j_1, \ldots, j_m\} = \{1, \ldots, m\}$ are identical. The aforementioned set of conditions may be empty in general case, i.e., $m = 0$. The S-production with the empty conditions set looking like

$$<s_0 \leftarrow , d>, \tag{4}$$

is called *S-axiom*.

Component d of S-production $\sigma = <q, d>$ is set of rules $\gamma \rightarrow \beta$ and is called here *variables declaration*.

Rule $\gamma \rightarrow \beta \in d$ is called *variable declaration of* γ, which defines $\mathrm{V}(\gamma, \beta)$ set of possible values (domain) of this variable:

$$\mathrm{V}(\gamma, \beta) = \{u \mid \beta \overset{*}{\underset{G}{\Longrightarrow}} u \,\&\, u \in V^*\}, \tag{5}$$

where G is CF-grammar, corresponding to metadatabase D. It is essential that there is one and only one variable declaration $\gamma \rightarrow \beta$ for every variable γ, having place in the body of S-production. S-production σ, which body $s_0 \leftarrow s_1, \ldots, s_m$ has no any variables, i.e.,

$$s_i \in V^*, \tag{6}$$

for all $i = 0, 1, \ldots, m$, is called *concrete S-production* (for short *CS-production*). CS-production:

$$<s_0 \leftarrow, d>, \quad (7)$$

such that $s_0 \in V^+, d = \{\varnothing\}$ is called *CS-axiom*.

S-production $\sigma = <q, d>$ defines set of CS-productions $\overline{\sigma}$ in the following way:

$$\overline{\sigma} = \bigcup_{\delta \in \overline{d}} <s_0[\delta] \leftarrow s_1[\delta], \ldots, s_m[\delta], \{\varnothing\}>, \quad (8)$$

$$\overline{d} = \bigcup_{V(u_1,\gamma_1,\beta_1)} \ldots \bigcup_{V(u_k,\gamma_k,\beta_k)} \{\{\gamma_1 \to u_1, \ldots, \gamma_k \to u_k\}\}, \quad (9)$$

where $s_i[\delta]$ are strings in terminal alphabet V, which are created by the replacement of variables $\gamma_1, \ldots, \gamma_k$ by strings $u_1, \ldots, u_k$ in the same alphabet, respectively; all occurrences of one and the same variable γ_i in all terms $s_0, s_1, \ldots, s_m$ are replaced by one and the same string u_i.

If

$$(\forall <w_0 \leftarrow w_1, \ldots, w_m, \{\varnothing\}> \in \overline{\sigma})$$
$$\{w_0, w_1, \ldots, w_m\} \subseteq L(G), \quad (10)$$

then S-production $\sigma \in S$ is *correct to* MDB D. This means that all terms of every CS-production from set $\overline{\sigma}$ are words of context-free language $L(G)$, and where G is context-free grammar with set of generating rules D.

APS KB $P = <S, D>$ is *correct*, if all S-productions $\sigma \in S$ are correct to metadatabase D.

Notion of the *APS KB extensional*, i.e., set of facts, which may be derived from the knowledge base by means of logical inference, is defined as follows.

Let $P = <S, D>$ be correct APS KB. Consider

$$\overline{S} = \bigcup_{\sigma \in S} \overline{\sigma}, \quad (11)$$

i.e., $\overline{S}$ is a set of all CS-productions defined by all S-productions of this knowledge base in the sense (8)–(9).

Define

$$W_{(0)} = \{w \mid <w \leftarrow, \{\varnothing\}> \in \overline{S}\}, \quad (12)$$

$$W_{(i+1)} = W_{(i)} \cup \left(\bigcup_{\substack{<w_0 \leftarrow w_1, \ldots, w_m, \{\varnothing\}> \in \overline{S} \\ w_1 \in W_{(i)} \\ \ldots \\ w_m \in W_{(i)}}} \{w_0\} \right), \quad (13)$$

and extensional $Ex(P)$of APS KB $P = <S, D>$, i.e., set of facts, defined by this knowledge base, is fixed point of the sequence $W_{(0)}, \ldots, W_{(i)}, W_{(i+1)}, \ldots$, which is in general case infinite:

$$Ex(P) = W_{(\infty)}. \quad (14)$$

Evidently, due to P correctness,

$$Ex(P) \subseteq L(G), \quad (15)$$

and $Ex(P)$ is finite, if their exists i such, that

$$W_{(i)} = W_{(i+1)} \tag{16}$$

If we consider the notion of extensional of APS KB from the linguistic point of view, then $Ex(P)$ is a language in alphabet V, which, according to (15), is sublanguage of $L(G)$. At the same time, from the point of view of knowledge engineering, $Ex(P)$ is the join of set of facts $w \in W_{(0)}$, which are known explicitly (they are called lower *ground facts*), and set of facts, which are derived from ground facts and/or another derived facts. As it is easy to see, set of ground facts $W_{(0)}$ is nothing else, than SDB, while S-productions with nonempty sets of conditions form *APS KB intensional*, providing the aforementioned inference.

Now we can define semantics of language of queries to APS KB.

Set-theoretical (S-) semantics of this language is similar to S-semantics of SDB query languages:

$$A = \overline{W} \cap I \tag{17}$$

where $\overline{W}$ is APS KB extensional and I is set of facts, which actuality check is the purpose of the query.

Here we shall use the simplest query language, being set of couples $<s, d>$, where s is term and d is its variable declaration.

Mathematical (M-) semantics of this language is based on (14) and the following evident equation:

$$I = Ex(<\{<s \leftarrow, d>\}, D>) \tag{18}$$

Here, $<\{<s \leftarrow, d>\}, D>$ is correct APS KB, which one-element set of S-productions—selection criterion—contains S-axiom, defining set of facts, which may belong to $Ex(P)$.

Example 1. Consider metadatabase from Example 2 of the chapter of this book, describing SSF. We shall add to it the following three rules:

$$<fact> \rightarrow SENSOR <i> AT <time> \; - \; <state>,$$
$$<i> \rightarrow <symbol> <symbol>,$$
$$<fact> \rightarrow SENSOR <i> LOCATED \text{ AT } AREA <name\ of\ area>.$$

Facts like $SENSOR\ NN\ AT$ 17.00 $-$ $NORMAL$ contain information about sensors, which gather and send to the fusion center the data about the state of the atmosphere in the surrounding area. Facts like *SENSOR NN LOCATED AT AREA Y...Y* contain information about sensors' location.

Consider APS knowledge base $P = <S, D>$, where MDB D was described higher, and S contains following S-productions:

$$\sigma_1 : <AREA\ a\ IS\ s\ AT\ t \leftarrow$$
$$SENSOR\ e\ AT\ t - s,$$
$$SENSOR\ e\ LOCATED\ AT\ AREA\ a,$$
$$\{a \rightarrow <name\ of\ area>, s \rightarrow <state>, e \rightarrow <i>,$$
$$t \rightarrow <time>\}>,$$

$$\sigma_2 : < SENSOR\ AX\ AT\ 16.00 - NORMAL \leftarrow, \{\varnothing\} >,$$

$$\sigma_3 : < SENSOR\ FY\ AT\ 11.30 - SMOKED \leftarrow, \{\varnothing\} >,$$

$$\sigma_4 : < SENSOR\ AX\ LOCATED\ AT\ AREA\ HIGHLANDS \leftarrow, \{\varnothing\} >,$$

$$\sigma_5 : < SENSOR\ FY\ LOCATED\ AT\ AREA\ GREEN\ VALLEY \leftarrow, \{\varnothing\} >.$$

As seen, $\sigma_2 - \sigma_5$ are concrete S-productions, and set $W_{(0)}$, corresponding to this APS KB, consists of ground facts, being headers of these CS-productions.

Query, whose purpose is to get information about smoked areas, may be as follows:

$$< AREA\ z\ IS\ SMOKED\ AT\ t, \{z \rightarrow < name\ of\ area >, t \rightarrow < time > \} >,$$

while query, whose purpose is to get information about sensor *FY* location, may look like

$$< SENSOR\ YF\ LOCATED\ AT\ AREA\ v,$$
$$\{v \rightarrow < name\ of\ area > \} >.$$

Evidently, this knowledge base extensional is

$$\{AREA\ HIGHLANDS\ IS\ AT\ NORMAL\ STATE\ AT\ 16.00,$$
$$AREA\ GREEN\ VALLEY\ IS\ SMOKED\ AT\ 11.30,$$
$$SENSOR\ AX\ AT\ 16.00 - NORMAL,$$
$$SENFOR\ FY\ AT\ 11.30 - SMOKED,$$
$$SENSOR\ AX\ LOCATED\ AT\ AREA\ HIGHLANDS,$$
$$SENSOR\ FY\ LOCATED\ AT\ AREA\ GREEN\ VALLEY\}.\blacksquare.$$

3. Operational semantics of APS

The background of operational (O-) semantics of the considered query language is the so-called S-unification, which is generalization of well-known unification, introduced by J. Robinson in [7] and used in the resolution procedures, developed for the first-order predicate logic. *(In fact, S-unification is a strict generalized definition of heuristically described unification, which, as it was shown in [1], is incorrect in the case of multiple occurrences of at least one variable to one of the unified terms.)*

We shall consider basic concept of S-unification and then describe operational semantics of used query language, corresponding to (11–18) and defining main features of the controlled logical inference of answers to queries to APS KB.

Let $< s, d >$ be a query and $< s_0^i, d_i >$ is couple, formed from S-production:

$$\sigma_i : < s_0^i \leftarrow s_1^i, \ldots, s_{m_i}^i, d_i > \tag{19}$$

As may be seen easily, the answer to $< s, d >$ may contain headers of the CS-productions, being concretizations of σ_i, i.e., belonging to set $\overline{\sigma}_i$, if

$$W_{\sigma_i}^{s,d} = Ex(< \{ < s \leftarrow, d > \}, D >) \cap$$
$$Ex(< \{ < s_0^i \leftarrow, d_i > \}, D >) \neq \{\varnothing\} \tag{20}$$

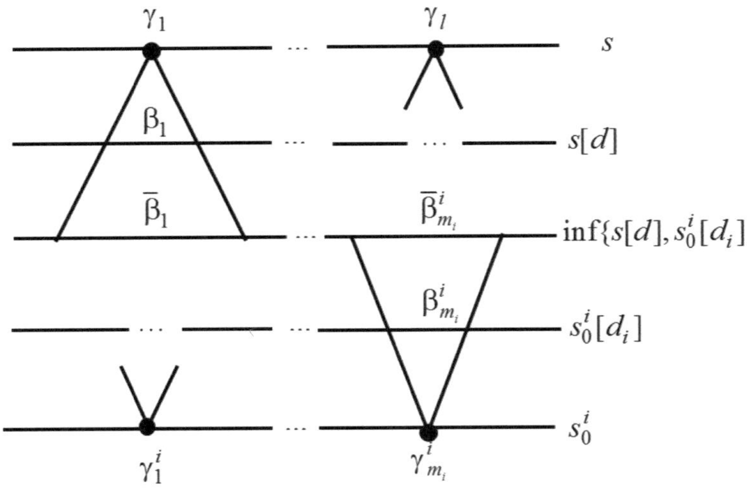

Figure 1.
Graphical illustration of S-unification.

This condition is equivalent to the existence of solution of word equation on context-free language (WECFL):

$$\left[s = s^i_o, d \cup d_i\right], \tag{21}$$

(for correctness, sets of variables of terms s and s^i_o would not intersect, that is, simply implemented by variables renaming, if this condition is not satisfied).

Result of WECFL (21) solution is the new declaration of variables of S-production σ_i, "concreted" in accordance with query $<s, d>$. This concretization will be denoted as $d_i\left(\frac{s^i_o}{s, d}\right)$; it is obtained by replacing variable declarations β^i_j of term s^i_0 by new ones $\overline{\beta}^i_j$, which are more informative ("concrete," "precise"), regarding initial declarations in d_i. This concretization is adequate to query $<s, d>$ in such a way, that CS-productions, whose headers may enter the answer to this query, may be created at the following steps of inference.

Logics of S-unification are illustrated in **Figure 1**, followed by Example 2.

Example 2. Consider query $<s, d>$, where

$$s = AREA\ e\ IS\ NORMAL\ AT\ t,$$
$$d = \{e \to <name\ of\ area>, t \to <time>\},$$

and S-production σ_1 from Example 10, where

$$s^1_0 = AREA\ a\ IS\ s\ AT\ t,$$
$$d_1 = \{a \to <name\ of\ area>, s \to <state>, e \to <i>, t \to <time>\}.$$

According to (21),

$$s^1_0[d_1] = AREA <name\ of\ area> IS <state> AT <time>,$$
$$s[d] = AREA <name\ of\ area> IS\ NORMAL\ AT <time>,$$
$$d_1\left(\frac{s^1_0}{s, d}\right) = d_1 - \{s \to <state>\} \cup \{s \to NORMAL\} =$$

$$= \{a \to <name\ of\ area>, s \to NORMAL, e \to <i>,$$

$$t \to <time>\}.\blacksquare$$

S-unification provides opportunity of answer derivation by controlled logical inference, which is implemented by the axiomatic system, which contains only three inference rules—top-down successful S-resolution, top-down unsuccessful S-resolution, and bottom-up S-resolution [1, 2]. Here we shall describe more natural and understandable **procedural representation** of the considered O-semantics, which fully corresponds to the aforementioned axiomatics.

This representation contains two recursively interconnected algorithmically defined functions.

The first of them Q has two input variables s and d, which values represent query $<s, d>$, and one output variable A, which value, returned after Q termination, is the answer to this query, as defined by (17).

The second function RB has also two input variables, φ and d, which values represent, respectively, the so-called residual body RB of S-production (subset of body of S-production, including terms, which until current call were not used for the derivation of new queries) and variable declaration, obtained after previous steps of logical inference. Function RB returns set $\mathcal{D}$of variable declarations, each corresponding to one element of answer to the query $<s, d>$, where $s \in \varphi$ is the next term, extracted from RB. By this, input variables declaration d is made "more precise."

Text of function Q is as follows:

1 Q : **function**(s, d) **returns** (A);
2 **variables**$((s, s_0)$ **term**, (d, d', δ, δ') **declaration**, φ **set of terms**,
3 A **set of words initial**$(\{\varnothing\}))$ **local**;
4 **variables**$(S$ **set of productions**, D **set of context − free rules**$)$ **global**;
5 **do** $<s_0, \varphi, d'> \in S$;
6 $\delta := d'\left(\frac{s_0}{s, d}\right)$;
7 **if** $\delta \neq \nabla$
8 **then do** $\delta' \in RB(\varphi, \delta)$;
9 $A : \cup\{s_0[\delta']\}$;
10 **end** δ';
11 **end** S;
12 **end** Q

As seen, lines 2–4 contain declarations of variables, used lower in the function body. It is essential that there are two variables declared as global, namely, S (set of S-productions) and D (metadatabase). Nevertheless the last one is not used in the body, it is presumed that MDB is used inside S-unification (line 6). The initial value of local variable A is an empty set.

The main part of Q is loop (lines 5–11) on S-productions, entering set S and represented in the form of triples $<s_0, \varphi, d'>$, where s_0 is header; φ is body of S-production, being set of terms; and d' is its variable declaration. The first operator inside loop (line 6) implements S-unification of query $<s, d>$ and couple $<s_0, d'>$. If this S-unification is successful (i.e., its result is not a special value ∇), then loop on elements of set of variable declarations, obtained by function RB, is executed (lines 8–10). The values of input variables of function RB are φ (set of terms, entering body of the current S-production) and δ (result of the aforementioned S-unification). Each δ', being set of variable declarations of the form $\gamma \to u$, where $u \in V^*$, is used for making facts $s_0[\delta']$ by substitution right parts of those declarations to term s_0, and all such facts are joined to the output set A (line 9).

Text of function *RB*, in turn, is as follows:

1 RB : **function**(φ, d) **returns** $(\mathcal{D})$;
2 **variables**(φ **set of terms**, d **declaration**, s **term**, w **fact**,
3 $\mathcal{D}$ **set of declarations initial**$(\{\varnothing\})$ **local**;
4 **variables**(D **set of context−free rules**) **global**;
5 if $\varphi = \{\varnothing\}$
6 **then** $\mathcal{D} := \{d\}$;
7 **else do** $s \in \varphi$;
8 **do** $w \in Q(s, d)$;
9 $\mathcal{D} : \cup RB\left(\varphi - \{s\}, d\left(\frac{s}{w, \{\varnothing\}}\right)\right)$;
10 **end** w;
11 **end** s;
12 **end** RB

Here lines 2–4, as in the text of *Q*, are declarations of variables, used in the *RB* body. *RB* execution begins by check, whether set φ is empty (line 5): if so, search of terms of S-production body is already finished (or, as partial case, it is empty because S-production is axiom), and the returned value is one-element set $\{d\}$, where d is the same, as input value. Otherwise loop on terms, entering residual body, is executed (lines 7–11). Every selected term s is used in query $<s, d>$ to KB $<S,D>$, which is processed by recursive call $Q\ (s,d)$, and loop on all facts, entering answer to this query, is executed (lines 8–10). The body of this loop contains single operator (line 9), providing accumulation of set $\mathcal{D}$ as a result of recursive call of the same function *RB* with the first input variable being $\varphi - \{s\}$(i.e., set of terms of the body of the processed S-production, reduced by extraction of the processed term s) and the second input variable value being $d\left(\frac{s}{w, \{\varnothing\}}\right)$ (i.e., S-production variables declaration, made "more precise" according to result of S-unification of query $<s, d>$ and fact w).

As it is easy to see, execution of loop $s \in \varphi$ provides generation of $n!$ sequences of terms of body of S-production, i.e., all possible permutations of these terms.

Let $Q(s, d)$ be finite result of function *Q* application to the query $<s, d>$ and APS KB $<S,D>$. There is theorem on the correctness of the described operational semantics of the augmented Post systems [2].

Theorem 1. $Q(s, d) = Ex(S, D) \cap Ex(<\{s \leftarrow\}, \{d\}>)$.

In fact, *Q* and *RB* collaborative operation provides top-down derivation of the set of CS-productions, involved in creation of non-ground facts, which enter the answer to the query $<s, d>$, in full accordance with bottom-up definition of M-semantics of APS. (Ground facts are simply selected by one-step application of *Q* and *RB*.)

As seen from this description, logical inference is implemented in a wavelike manner. "Direct waves," including generated queries, which contain "more and more concrete" variables declarations, are "spreading" until S-axioms are reached. This leads to the creation of CS-productions, and "back waves" start until a new set of facts, being answer to some intermediate (or, finally, initial) query, is assembled. Both direct and back waves meet one another and "interfere," producing new queries. As may be seen, APS operational semantics corresponds to the *universal*

DOI: http://dx.doi.org/10.5772/intechopen.86207

mode of inference (all facts, being goal of the query, are derived), unlike *existential* mode (anyone fact from the possibly multifact set), which is inherent to known resolution procedures. By this, full accordance of APS O-semantics to its S- and M-semantics is achieved.

4. Procedural connection to APS KB

One of the most important features of any knowledge representation model are embedded tools of the procedural connection, providing interaction of "rigid" (non-updated by knowledge engineers) software modules with inference process and data exchange between inference engine and the aforementioned modules.

Background of the **procedural connection to APS knowledge bases** are *program (P-) productions*.

P-production is couple $< s_0 \Leftarrow p, d >$, where term s_0 is header and set d variable declaration; s_0 and d are declarative component of the procedural connection. Symbol $'' \Leftarrow ''$ is the divider, distinct from $'' \leftarrow ''$, and p is the name of the connected program (software module), which has its own extensional $W_p \subseteq Ex$ $(<\{< s_0 \leftarrow, d >\}, D >) \subseteq L(G)$.

Couple $< s_0, d >$ is called *cover of the connected program* p. In general case one and the same program may be connected to APS KB by several P-productions with different covers.

Example 3. Program named MULT for multiplication of integer numbers may be connected to knowledge base by P-production:

$$< a * b = c \Leftarrow \text{MULT}, \{a \rightarrow \text{<integer>}, b \rightarrow \text{<integer>}, c \rightarrow \text{<integer>}\} >.$$

MULT extensional is an infinite set containing strings like $1*1=1$, $0*5=0$, $311*-1=-311$, etc.■

Let Π be the set of names of programs, connected to APS KB. The extensional of this KB will differ from (12) to (14) by the only feature; instead of (12) the following is used:

$$W_{(0)} = \left(\bigcup_{p \in \Pi} W_p \right) \cup \{ w | \{ w \leftarrow, \{\varphi\} > \in \overline{S} \}, \qquad (22)$$

i.e., $W_{(0)}$ is a join of extensionals of all connected programs and set of ground facts, being heads of CS-axioms. Until general case will be discussed, we shall consider programs with the so-called static extentionals, whose basic feature is $W_p = \text{const}$ for all period of query processing.

Concerning operational semantics, it is sufficient to extend function Q by operators, providing call of the "perspective" connected programs via unified application programs interface and joining resulting sets to the accumulated answers. The extension is as follows:

```
do <s0, p, d'> ∈ S;
    δ:=d'(s0 / s, d);
    if δ ≠ ∇
        then A : ∪p(s0, δ);
end S;
```

As seen, after search on S-production set (lines 5–11), similar search on P-production set would be executed (lines 12–16). If S-unification of query $<s, d>$ and program p cover $<s_0, d'>$ is successful, program call is executed, and its result —set of words, denoted $p(s_0, \delta)$—is joined to the accumulated set A. This fully corresponds to (22).

As seen, search on the set S is nondeterministic (there is no any predefined order on this set, except S-productions processing before P-productions; but in general case, this ordering, of course, is not mandatory). This obstacle creates some difficulties in implementation of calls of the connected programs, when input data are less informative, than it is necessary for computation (e.g., MULT$(a * b = c, \{a \rightarrow <integer>, b \rightarrow 4, c \rightarrow <integer>\})$. By this, result of such call may be infinite set, even if there would be an opportunity to implement this operation using tools for N-facts processing, developed in [1–4].

To avoid such difficulty, it would be reasonable to implant to S- and P-productions some information, which may cut off "dangerous" branches.

There is a simple tool for logical inference control within APS knowledge representation. Namely, variables, which the declarations after S-unification would have right parts, consisting only of terminal symbols (i.e., there would be no incomplete information, associated with non-terminals, inside these declarations), are marked by point over arrow in the initial descriptions, having place in P- and S-productions. Such variables are called *complete*. In example 3 declarations of variables a and b would be $a \dot{\rightarrow} <\text{integer}>$, $b \dot{\rightarrow} <\text{integer}>$, so query

$$<c * e = f, \{c \rightarrow 1, e \rightarrow 4, f \rightarrow <integer>\}>$$

while inference will result in program MULT call, while query $<x * a = s, \{x \rightarrow <integer>, a \rightarrow 135, s \rightarrow <integer>\}>$ will not, because of information incompleteness of the declaration of x, which is a complete variable. Negative result of S-unification in the case of such incompleteness is, as higher, denoted by symbol ∇.

Let us consider now the general case, where extensionals of programs, connected to APS KB, are not static, i.e., may vary while answer derivation. The simplest examples of such kind of programs are database management systems.

To connect DBMS, operating key-addressed databases, it is sufficient to join to APS KB P-production:

$$<?k = d \Leftarrow DBMS, \{k \dot{\rightarrow} <key>, d \rightarrow <data>\}>$$

for queries (variable k, denoting key, is complete), as well as

$$<k{:=}d \Leftarrow DBMS, \{k \dot{\rightarrow} <key>, \dot{d} <data>\}>$$

for update operations (both variables k and d, corresponding to key and data, are complete).

DBMS, operating databases with symmetric access, may be connected in a following way. In order to minimize redundant search while queries processing, some substrings of facts may be declared indexed (used for creation and maintenance of dynamic search trees, as it is described in [2, 3]). So some covers of the relational DBMS, corresponding to some subsets of DB, may contain complete variables, with declarations like $\gamma\dot{u}$, where $\rightarrow u$ defines domain of the indexed attribute of the relation. Covers, providing inclusion to SDB, contain only complete variables. (If DBMS operate SDB with incomplete information, any variable may be incomplete.)

5. Flow-processing APS

Described approach to the procedural connection makes easy integration of the distributed hardware components, linked by the computer network, into a single system with the single operation process. This may be done by procedural connection of the corresponding hardware drivers and implementation of the so-called flow-processing augmented Post systems.

FAPS KB includes along with S- and P-productions also the so-called flow (F-) productions having the form of

$$< s_0 \rightarrow s_1, \dots, s_m, s_{m+1}, d > \tag{23}$$

where terms s_0, called activator, and s_{m+1}, called actor, contain only complete variables and symbol "→" (inverse to "←", used in S-production) divides activator and body, including, along with actor, conditions $s_1, \dots, s_m$. This structure is rather usual as well as operational semantics of F-productions, which is defined by function *F*, similar to *Q* and *RB*, and interconnected with them into integrated algorithmics.

We assume that F-productions are represented in the knowledge base as couples $< s_0, \varphi, s, d >$, where s_0 is activator, φ is set of terms-conditions, and s is actor, while d is variable declaration. By this, APS KB contains three types of objects, corresponding to three types of productions:

1. $< s_0, \varphi, d >$, whereφ is set of terms (S-productions);

2. $< s_0, p, d >$, where p is string, being name of the connected program (P-productions);

3. $< s_0, \varphi, s, d >$ (F-productions).

All these objects are accumulated to set *S*, which, along with metadatabase *D*, present in the bodies of functions *Q*, *RB*, and *F* as global variable.

Function *F* is as follows:

1 F : **function**(w);

2 **variables**$((s_0, s')$ **term**, (d, δ, δ') **declaration**, w **word**, φ **set of terms**) **local**;

3 **variables**(S **set of productions**, D **set of context−free rules**) **global**;

4 **do** $< s_0, \varphi, s, d > \in S$;

5 $\delta := d'\left(\dfrac{s_0}{w, \{\varnothing\}}\right)$;

6 **if** $\delta \neq \nabla$

7 **then do** $\delta' \in RB(\varphi, \delta)$;

8 $F(s[\delta'])$;

9 **end** δ';

10 **end** S;

11 **terminate**;

12 **end** F

As seen, the search on F-productions set (lines 4–10) provides selection of such F-productions, which are activated by input message w. The set of conditions of every activated F-production is interpreted as residual body of S-production (lines 7–9). Every variable declaration δ', obtained as a result of this interpretation, is used for the creation of new message by substitution of δ' to actor s. This message is used as input value for function F recursive application. So the wave of messages, triggered by initial message, is generated, modeling well-known *blackboard architecture*. This wave propagation of any programs (including DBMS, software/hardware drivers, providing activation and operation of various devices, as well as networking middleware) may be applied. Operator **terminate** (line 11) stops F execution, so no return to the parent call is performed.

Various practice-oriented dialects of APS, providing various effective technologies of non-procedural programming (multiactivating, triggers, terms with negators, decision tables, etc.), were developed; also described operational semantics of APS and its extensions were redefined for highly parallel hardware environment [1, 2]. Taking into account nondeterministic nature of operational semantics of APS family, some additional tools for logical inference control were introduced in [1, 2].

Let us consider now the main features of APS application to the most advanced areas—Big Data, Internet of Things, cyberphysical industry, and cybersecurity—forming the future digital economy information infrastructure (DEII) [13, 14]. In all aforementioned applications, APS knowledge representation plays interconnecting and flexifying role, providing fast integration of various heterogeneous systems and fast adaptation of their operation logic to the highly volatile environment. In fact, APS simplify to the maximally possible level the most complicated problem of *interoperability* of any a priori developed systems; APS KB is nothing but "glue," which in the simplest regular way integrates them together.

6. Implementational issues and applications of APS

DEII is the result of integration of three basic paradigms: network-centricity, Big Data, and Internet of Things (**Figure 2**).

Network-centricity provides interconnection of any subjects of digital economy (DE), integrating human society and technosphere into global technosocium, operating as a global megasystem for the mankind wealth and prosperity. Big Data provides storage of great amounts of data from multiple heterogeneous sources and

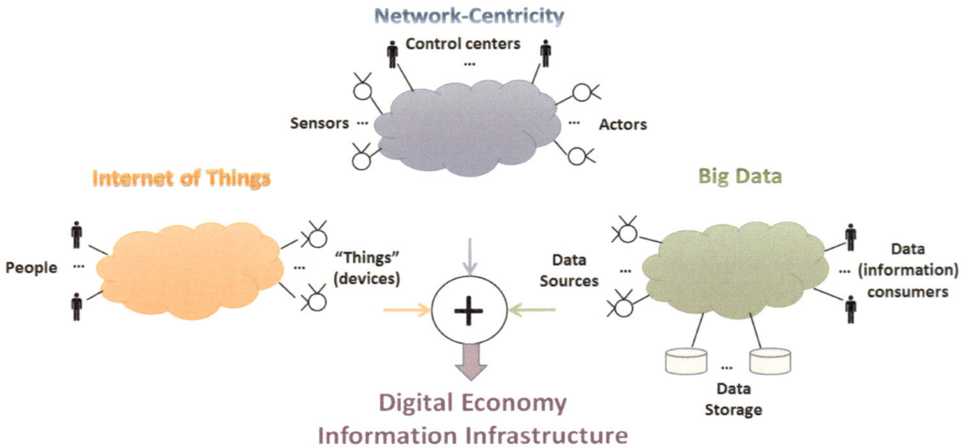

Figure 2.
Basic paradigms of digital economy information infrastructure.

use of these data for rational everyday operation of the aforementioned megasystem and its permanent improvement. The Internet of Things, whose more precise and correct name would be Internet of Devices, provides creation, development, and operation of the aforementioned future global technosphere, including cyberphysical industry, based on additive manufacturing technologies, deeply robotized smart logistics, smart energy generation and delivery systems, and life resources infrastructure (food, water, living houses, etc.), as well as digital banking and finance infrastructure. All these segments, containing many billions of interconnected devices, provide implementation of such ambitious initiatives as Smart City and Smart Nation, inevitably leading to the Smart World as a well-understood and achievable goal of mankind evolution.

The most complicated problem to be solved for DEII creation and development is providing its flexibility, i.e., rapid correction of operation logic of various DEII elements and sets of elements to the constantly occurring changes of the environment, as well as to changes in our knowledge about nature, human society, and technosphere. It is evident that sufficient flexibility of DEII may be achieved only on the background of knowledge engineering, leading to the knowledge-based digital economy.

As shown in **Figure 3**, every subject of the DEII (no matter, human, or device), being associated with unique address of the address space of global information infrastructure, is supported by local knowledge base (LKB), applied by knowledge interpreter-corrector (KIC) for processing of input information flow, as well as local database, used and updated while aforementioned processing. Here LKB may contain not only "soft" component, i.e., rules, defining logic of input messages processing, but also "firm" component, i.e., "rigid" (nonmodified) software modules and systems (up to DBMS clients and servers), connected to LKB by the common interface, and called while rules interpretation (their sets are marked CP, e.g. Connected Programs, with lower indices).

The described approach may be effectively implemented, if there would be some unified data item, providing unified representation of operation logic of any system, joining any set of DE subjects.

The possible practical outcome of the SSF is the creation of toolkit, providing the described higher approach to DEII implementation upon **string as the aforementioned unified data item**. This outcome in the integrated form is placed in

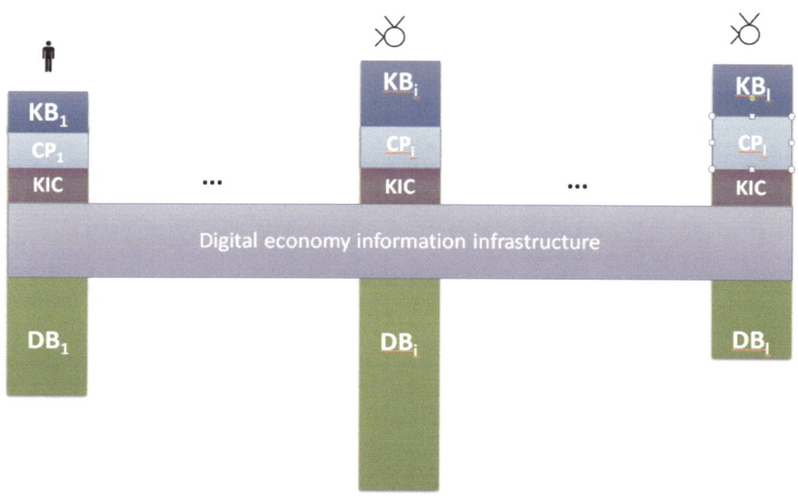

Figure 3.
Linearized representation of the knowledge-based digital economy information infrastructure.

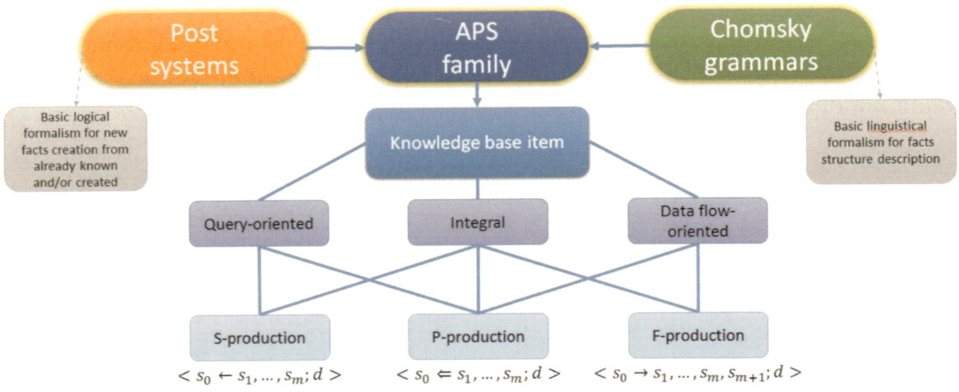

Figure 4.
Integrated representation of APS family of knowledge representation models.

Figure 4, where APS family of string-operation knowledge representation models is explicated.

The operational semantics of the simplest practically used model from this family, whose background is FAPS, is shown in **Figure 5**.

As seen, input messages, which are transported to the local subject of DEII by means of network infrastructure, are recorded on the external blackboard (in fact, there are two blackboards used: external, for messages from the external sources, and internal, for messages generated while current external message processing). Every next message (string) w is navigated by special associative index (binary/ ternary tree) to the activators s_o^i. Selected activators are used for solving the corresponding WECFL (to simplify variable declarations and accelerate messages parsing, the so-called ultragrammars were introduced in [1, 2]), and if solutions do exist, they are transferred to the rest part of the F-production, in which the terms may provide *access to databases with symmetric access (DBSA) or key-addressed databases (KADB)* SQL-like or NoSQL queries and updates; *check database integrity* before updates (by applying constraints defined by the corresponding sets of productions); *activate devices* by their drivers, connected to LKB by proper P-productions;

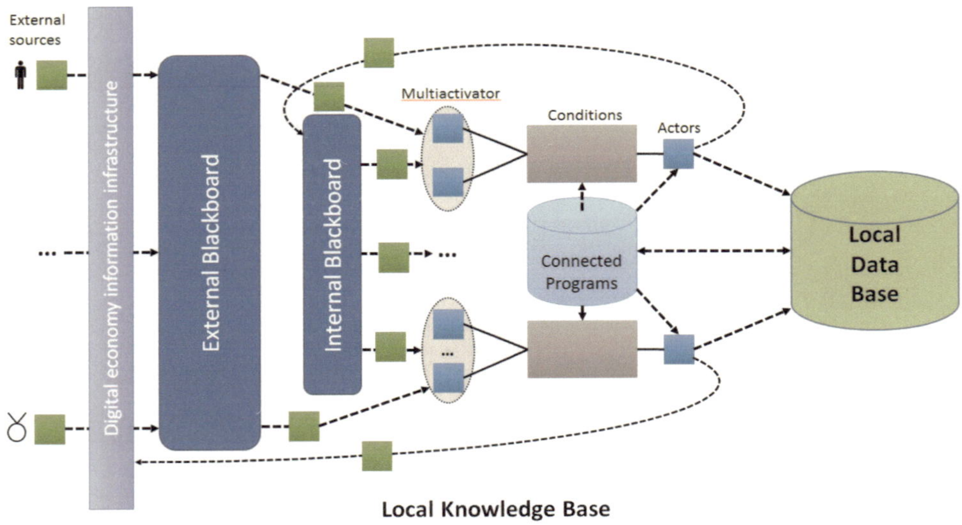

Figure 5.
Basic operation semantics of multiactivated flow APS.

perform some nontrivial *processing by connected "firm" software products*; and also *send messages*, which are created strings to the internal blackboard or, by lower-level communication software, to another local subject (i.e., their external blackboards).

As seen, described dialect of APS family provides flexible implementation of operations, necessary for Big Data and Internet of Things, as well as its cyberphysical manufacturing segment, i.e., industrial Internet of Things. In the simplest case, strings, being transferred to 3D printers by their driver's calls, may be STL files containing layer descriptions of the printed material objects [15].

The main feature, which is, in fact, core for all APS family, is "additivity" of the local knowledge bases. Namely, the occurrence of any new device or human, generating earlier unknown and unprocessed messages, is supported by addition to LKB of subjects (again devices or humans), communicating with source of such messages, new elements with activators, providing initiation of their processing. Such occurrence results, finally, in creation and sending new messages to another subjects. Such "additivity" may be called "vertical" (new types of messages are supported by the addition of new F-productions with new activators to LKB). However, horizontal "additivity" is also supported, when new F-productions with proper activators are added for internal messages, which are sent by some modules of LKB while processing external messages. By this, internal messages processing becomes "deeper." As seen, FAPS provide sufficiently regular and refined technology of the distributed system software development and debugging.

Let us underline that unlike well-known tools from the object programming area, which also support interconnection between modules by messages and use blackboard for such processing, in the case of flow APS, new message receivers are not known, and thus, there is no opportunity for its determination in the text of the program module. By this, APS-based knowledge engineering approach to software making principally differs from object programming as well as any other approaches from the more or less procedural programming.

Let us pay some attention to such important area of flow APS application as cybersecurity, which is critical for DEII normal operation.

Dynamically extended spectrum and complexity of cyberattacks, implementing today advanced persistent threats (APT), have led to the necessity of development of more and more sophisticated tools for early recognition and prevention of APT. The most efficient of such tools are based on the security information and event management (SIEM) paradigm [16–19], whose background technologies are deep packets inspection/deep packets processing (DPI/DPP) [20, 21] and data leakage prevention (DLP) [22, 23]. The first operates flows of bit packets, providing their fusion in order to recognize elaborately covered signatures of known cyberattacks, while the second operates usually traffic from the application level of the OSI model. However, both operate strings (no matter, bit, symbol, or combined), and by this reason the described higher FAPS are the perfect tool for the SIEM implementation, especially at security operation centers (SOC), providing fusion of the primary data, passing to SOC from large amount of software and hardware traffic sensors from the security perimeter of the protected system. (Such sensors may be also effectively implemented on the FAPS background.) It is very important that SIEM is solidly based on three pillars: Big Data, which are stored fragments of the network traffic, caught at many network links in a 24×7 regime; data mining; and knowledge engineering. This area is the hardest known case of data mining and knowledge-based technologies application: cybersecurity knowledge engineers, by analyzing accumulated enormous volumes of primary data items, must quickly understand signatures of the prepared or already performed cyberattacks, developed by the smartest people from the global cybercrime and states special services teams; and rapidly correct LKBs or extend them by new sets of productions, providing early recognition and neutralizing such attacks. This combination of very large volumes of data and knowledge with hard real-time regime of SOC operation (e.g., SOC of the Russia largest financial group Sberbank neutralizes daily over 14,000 cyberattacks from all over the world [24]) makes cybersecurity application of data and knowledge engineering the most important from both practical and theoretical points of view.

7. Conclusion

The described "Set of Strings" Framework is based on integration of three research areas—knowledge/data engineering, theory of grammars, and information theory—which until now are very close but rather isolated from one another. Synergetic effect, which may be the main result of such integration, would be useful for the development of the unified theory, necessary for convergence of Big Data, data mining, artificial intelligence, and Internet of Things. Such convergence is extremely needed for solution of all spectrum of problems of digital economy.

The most valuable directions of the SSF future development may be the following:

1. Metainference and machine learning in APS KB.

2. N-facts concretization in SDB with incomplete information (SDBI) and APS KB, using functional dependencies, associated usually with the relational data model.

3. Contradictory SDBI and APS KB management.

4. SDBI and APS KB with metadatabases, describing two- and three-dimensional objects.

5. SSF ideology and techniques transfer to the numeric data with the help of the SSF-associated theory of recursive multisets and multiset grammars/metagrammars, developed for the solution of problems of planning and scheduling in digital economy [25–28].

6. Hardware implementation of key elements of SSF in the data flow and other high-parallel computer environments.

The author is highly interested in the cooperation with computer scientists and engineers, who may spend some research effort participating in the development of the listed directions and the SSF at all.

Acknowledgements

The author expresses gratitude to the editor for useful advices. The author is grateful to Prof. Noam Chomsky for words of support to the development of the "Set of Strings" Framework. The author is also thankful to Prof. Jeffrey Ullman for useful remarks on the current state of the theory of grammars.

Author details

Igor Sheremet
Financial University under the Government of Russian Federation, Moscow, Russia

*Address all correspondence to: sheremet@rfbr.ru

References

[1] Sheremet IA. Intelligent Software Environments for Information Processing Systems. Moscow: Nauka; 1994. p. 544 (In Russian)

[2] Sheremet IA. Augmented Post Systems: The Mathematical Framework for Data and Knowledge Engineering in Network-Centric Environment. Berlin: EANS; 2013. p. 395

[3] Sheremet I. Data and knowledge bases with incomplete information in a "Set of Strings" framework. International Journal of Engineering and Applied Sciences. 2016;**3**(3):90-103

[4] Sheremet IA. Augmented Post systems: String-operating knowledge representation for Big Data and Internet of Things applications. Geoinformatics Research Papers. 2017;**5**:BS1002. DOI: 10.2205/CODATA 2017

[5] Chomsky N. Syntactic Structures. 2nd ed. Berlin–New York: Mouton de Gruyter; 2002. p. 117

[6] Post EL. Formal reductions of the general combinatorial decision problem. American Journal of Mathematics. 1943; **65**:197-215

[7] Robinson JA. A machine-oriented logic based on the resolution principle. Journal of the ACM. 1965;**12**:23-41

[8] Kowalski RA. Algorithm = Logic + Control. Communications of the ACM. 1979;**22**(7):424-436

[9] Nilsson U, Maluszynski J. Logic, Programming and Prolog. 2nd ed. John Wiley & Sons; 2000. p. 282

[10] Cali A, Gottlob G, Lukasiewicz T. A general datalog-based framework for tructable query answering over ontologies. Journal of Web Semantics. 2012;**14**(1):57-83

[11] Miller D, Nadathur G. Programming with Higher-Order Logic. Cambridge University Press; 2012. p. 322

[12] Gallaire H, Minker J, Nicolas J-M. Logic and databases: A deductive approach. ACM Computing Surveys. 1984;**16**(2):153-185

[13] The Global Information Technology Report 2016. In: Baller S, Dutta S, Lanvin B, eds. Innovating in the Digital Economy. INSEAD, 2016. p. 62

[14] Lee J, Bagheri B, Hung-An. A cyber-physical systems architecture for industry 4.0: Based manufacturing systems. Manufacturing Letters. 2015;**3**: 18-23. DOI: 10.1016/j.mfglet/2014.12.001

[15] Eyers DR, Potter AT. Industrial additive manufacturing: A manufacturing systems perspective. Computers in Industry. 2017;**92-93**: 208-218. DOI: 10.1016/j.compind.2017.08.002

[16] Miller DR, Harris S, Harper A, Vandyke S. Security Information and Event Management (SIEM) Implementation. McGraw Hill; 2010. p. 464

[17] Miller DR, Harris S, Harper A, Vandyke S, Blask C. Security Information and Event Management (SIEM) Implementation. McGraw Hill; 2011. p. 465

[18] Sweeny J. Creating Your Own SIEM and Incident Response Toolkit Using Open Source Tools. SANS; 2011. p. 24. Available at: //sans.org/

[19] Limacher M, Kurz U. Security operation centre in action. CryptoMagazine. 2013;(3):10-12

[20] Fuchs C. Implications of Deep Packet Inspection (DPI) Internet

Surveillance for Society. Research Paper. Uppsala University. 2012. p. 125

[21] Bass T. Intrusion detection systems and multisensor data fusion. Communications of the ACM. 2000; **43**(4):99-105

[22] Papadimitriou P, Garcia-Molina H. Data leakage detection. IEEE Transactions on Knowledge and Data Engineering. 2011;**23**:51-63

[23] Shabtai I, Elovici Y, Rokach L. A Survey of Data Leakage Detection and Prevention Solutions. NY: Springer-Verlag; 2012. p. 92

[24] Sheremet IA. Digital economy and cybersecurity of its financial sector. Proceedings of Free Economical Society. 2018;**210**:23-34 (In Russian)

[25] Sheremet IA. Recursive Multisets and Their Applications. Berlin: NG Verlag; 2011. p. 249

[26] Sheremet IA. Multiset approach to the estimation of consequences of natural disaster impacts on industrial systems. Geoinformatics Research Papers. Sochi 2016;**4**:BS4002, DOI: 10.2205/2016 BS01 .

[27] Sheremet IA. Multiset analysis of consequences of natural disasters impacts on large-scale industrial systems. Data Science Journal. 2018; **17**(4):1-17. DOI: 10.5334/dsj-2018-004

[28] Gvishiani AD, Roberts FS, Sheremet IA. On the assessment of sustainability of distributed sociotechnical systems to natural disasters. Russian Journal of Earth Sciences. 2018;**18**:ES4004. DOI: 10.2205/2018ES000627

Chapter 12

Serialization in Object-Oriented Programming Languages

Konrad Grochowski, Michał Breiter and Robert Nowak

Abstract

This chapter depicts the process of converting object state into a format that can be transmitted or stored in currently used object-oriented programming languages. This process is called serialization (marshaling); the opposite is called deserialization (unmarshalling) processes. It is a low-level technique, and several technical issues should be considered like endianness, size of memory representation, representation of numbers, object references, recursive object connections and others. In this chapter we discuss these issues and give them solutions. We also include a short review of tools currently used, and we showed that meeting all requirements is not possible. Finally, we presented a new C++ library that supports forward compatibility.

Keywords: serialization, marshaling, deserialization, unmarshalling, archive, forward compatibility

1. Introduction

Serialization or *marshaling* is the process of converting object state into a format that can be transmitted or stored. The serialization changes the object state into series of bits. The object state could be reconstructed later in the opposite process, called *deserialization* or *unmarshalling*. The reconstructed object is a semantically identical clone to the original object. The object after serialization is called *archive*. Serialization is a low-level technique that violates encapsulation and breaks the opacity of an abstract data type.

Many popular programming languages have serialization support included in the language core or in the standard library. In the optimal situation serialization (and deserialization), the object does not require any additional development and code. In other programming languages, serialization is supported partially, and the usage needs some support in more advanced cases. In particular, C++ standard library contains stream representation as well as conversions between a binary or text streams and built-in data types. However there is no support for more advanced constructions like pointers, references, variants, collections and objects. Developers are required to rely on additional libraries or to manually write serialization code.

Manual creation of code to write and read object is time-consuming and liable to mistakes. This is the reason why libraries supporting serialization are provided. The other reason for use of external tool to serialization is their support to exchange of information between modules developed in different programming languages or executed on systems with different architectures. This functionality requires language-independent description of the data structure. Serialization tools that allow this data exchange are referred to as *portable*.

The portable serialization is not a straightforward process because:

- Different processor architectures (big-endian, little-endian) lead to a different binary representation of numbers and other objects, which potentially hinders portability.
- Objects accessed by reference should be properly restored in terms of inheritance and multi-base inheritance.
- The support of variants (unions in C), where the same memory buffer is used to store different objects.
- Complex structure, where single object could be referenced multiple times by various pointers and references, needs to be restored properly.
- Various object collections should be supported, including lists and dictionaries.

The good serialization support tools give possibility to choose the so-called archive type, i.e. the format used by serialization. *Archive medium* is a name for file or stream. Serialization archive formats can be divided into two main categories: text-based (stream of text characters) and binary format (stream of bytes) [1]. Typical examples of text-based formats include raw text format, JavaScript Object Notation (JSON) and Extensible Markup Language (XML). Binary formats are more implementation-dependent and are not so standardized. The stream of bytes is mostly memory- and time-efficient; therefore the serialized buffer is the smallest and usually fastest to marshall and deserialize; however the buffer is unreadable to developers and most susceptible for portability issues. Text formats are human-readable, which allows developers to perform manual inspections of archives and usually means easier portability, even across languages, but converting objects to text is usually more time-consuming, and memory footprint highly depends on serialized data and object structure. Raw text format is the most memory-efficient text format and is readable for developers if object structure is simple. JSON is a text format that supports tree-like object structures and allows simple validation; XML is also a text format that supports tree-like object structures [2]; moreover it is self-descriptive and allows data validation; however it takes a lot of memory.

As a result *portability* is used in at least two contexts:

- Portability across machines with different architecture but inside the same language or framework (implementation can rely on language-specific solutions, i.e. default character encoding)
- Portability across various languages and frameworks (usually that includes portability across various platforms), which faces various issues and often needs to introduce various constraints for possible serializable structures

During the software development process, it may be necessary to change the object structure being serialized. If the newer version of software is able to read data saved by the older version, the serialization mechanism has *backward compatibility*. If, additionally, the older version of software is able to read data saved by newer version, the serialization mechanism has *forward compatibility*.

Chapter 2 describes in more details common issues faced by serialization tool developer, followed by examples of existing solutions in Chapter 3. Chapter 4 presents proposed extension to one of the existing tools targeting C++ language. Conclusions are included in Chapter 5.

2. Technical issues for serialization

While designing serialization library, developers face various technical issues like different binary representation of numbers, different encoding of letters, object references, and inheritance. Those issues become especially troublesome when trying to create portable archive. Some of the most common issues related to creation of portable binary archives are depicted below together with possible solutions.

2.1 Numbers

There are two main issues with making archive portable between platforms when storing numbers—endianness and size of memory representation.

Two incompatible formats are in common use to represent larger than 1 byte numerical values when stored into memory: a big-endian, where the most significant byte (i.e. byte containing the most significant bit) is stored first (has the lowest address), and little-endian, where the most significant byte is stored last, has the highest address, as depicted in **Figure 1**. Big-endian is the most common format in data networking (e.g. IPv4, IPv6, TCP and UDP are transmitted in big-endian order), and little-endian is popular for microprocessors (Intel x86 and successors are little-endian, but Motorola 68,000 store numbers in big-endian; PowerPC and ARM support both).

Additionally various languages differently define their 'basic integer type'. For example, languages like Java or C# define `int`. as 32-bit variable, disregarding execution platform architecture. C and C++ use the platform-dependent definition of `int`—it has to be at least 16-bit, usually 32-bit on modern architectures, but can be anything bigger. Python (since version 3) can serve as the most extreme example, where built-in type `int`. is defined as *unbounded*. As a result saving types such as C++'s `std::size_t` or `long` directly, without additional size information, may produce data which may not be readable on other platforms.

Some common solutions include number size as part of serialized data or user forced to explicitly state size of data during serialization and deserialization, for example, by using a method named `writeInt16` or by using types like C++'s `std::uint64_t`. Potentially a more portable but possible less-efficient way is based on variable-length integer encoding, as described in the next section as solution for strings and array length encoding.

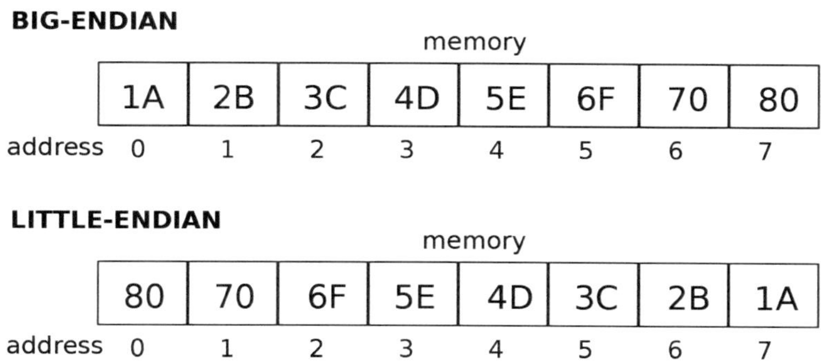

Figure 1.
Representation of `0x1A2B3C4D5E6F7080` *in big-endian and little-endian.*

Floating-point number is more standardized across platforms. The IEEE Standard for Floating-Point Arithmetic (IEEE-754) [3] describes, among others, binary representation of floating-point numbers. It is implemented by most of platforms used today. But that does not necessary solve floating-point numbers' portability issues—for example, the first version of the standard allowed different representations of quiet Not a Number (NaN). For example, Intel x86 and MIPS architecture can use different binary representations for it. Most serialization libraries support only platforms supporting IEEE-754 standard but still need to include their own representation of those NaNs and provide proper conversions.

Handling *long double* type, which is present on some platforms, is especially difficult as it is not standardized, and 64-bit, 80-bit or 128-bit types can be found. Usually portable serialization either does not support that type or defines it by itself, requiring proper conversions to be executed during serialization.

IEEE-754 does not specify endianness used to represent floating-point numbers; in most implementations endianness of floating-point numbers is assumed to be the same as endianness of integers.

2.2 String serialization

The biggest problems with string serialization lie in character encoding and variable-length optimization. Some languages or libraries (C#, Java, C++ Qt) force default encoding of character string in memory (usually from UTF [4] encoding family); others (C, C++) rely on platform or user settings. Serialization tool either forces its chosen encoding, which might lead to both time- and memory-inefficient encoding, or leaves character encoding unchanged and requires users to handle the issue using other means. The latter is usually chosen, as sheer amount of possible combinations of available encoding on various platforms is just enormous.

In most languages character strings can have variable length. This forces serialization process to store length somehow, which raises the question on the type used to store length. Choosing some arbitrary large integer type might be excessive for short strings; choosing too small type might result in problems with serializing huge data chunks. A good solution, which trades some processing time for memory usage, is to use variable-length integer encoding—then, for example, for small arrays, the size of the array is stored using only 1 byte.

The example of variable-length integer encoding is variable-length quantity (VLQ), where 8-bit bytes are used for storing integer and 7-bit types starting from the lowest significant bit are used for coding value, while the most significant bit is used to mark the next byte as part of the encoded integer. The examples are shown in **Figure 2**.

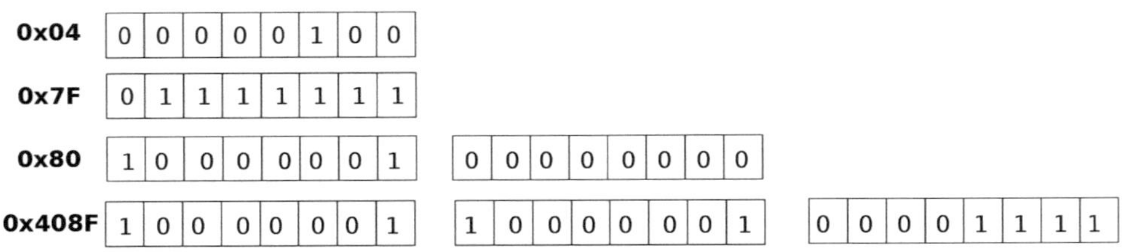

Figure 2.
Unsigned integer numbers variable-length encoded.

2.3 Object representation in memory

Potentially the fastest and easiest way to serialize an object would be to copy contents of memory where that object is stored. Even putting endianness issues aside and focusing on objects composed of only basic types (without pointers or references), this approach just cannot work in a portable way, due to differences in how an object is represented in memory. For efficiency some platforms might enforce specific memory alignment of fields, which introduces padding—empty 'unused'bytes between fields. Various platforms can have distinct memory alignments, which in turn can make the same object occupy different amounts of bytes on other systems. Some languages, like C++, give the user partial control over the object's memory layout, but even those features would not help in creating fully portable object representation.

2.4 Constant members and enumeration values

Serialization is a low-level technique, which violates encapsulation and breaks the opacity of an abstract data type. Object deserialization from this point of view is object creation using stream of bytes representing object state. Therefore the deserialization (unmarshalling) ignores constness, for example, by applying `const_cast` in C++. Users should either avoid serializing objects with constant values or provide proper constructors. Otherwise encapsulation has to be broken by serialization tool.

Popular languages provide 'enumeration types'—list of constants. Usually enumeration values are serialized using their integer representation. Serialization process has to ensure cohesion of those integer representations. The deserialization usually requires to create enumeration value from its integer representation; therefore, as for the constant, it acts as a low-level technique that breaks the rules of encapsulation.

2.5 Object references

A proper complete serialization should follow all references used in the object. Otherwise deserialized object would not be a semantic copy of its source. This means serializing or deserializing pointer or reference should act on pointed data. In other words serialization should do deep pointer (references) save and restore.

This is simple only when each reference is used only once—when object connections are tree-like. Then serialization could just 'flatten' such structure, serializing referenced objects as parts of holder objects. Problem arises, when multiple references point to the same object and uniqueness of referenced object is important for proper semantics of the whole structure.

Consider situation depicted in **Figure 3**, where object B is a 'shallow copy' of object A (it points to the same data—object C). In such situations only one copy of data should be saved into the stream, as depicted in **Figure 4**. To achieve that serialization tool must introduce some portable object identifiers and reference tracing. The latter might be difficult, depending on the language and its features used—for example, array of elements in C++, where each of the elements could be referenced by its direct address.

The issue becomes even more difficult in case of recursive object connections, depicted in **Figure 5**. The serialization tool has to detect such structures using reference tracing.

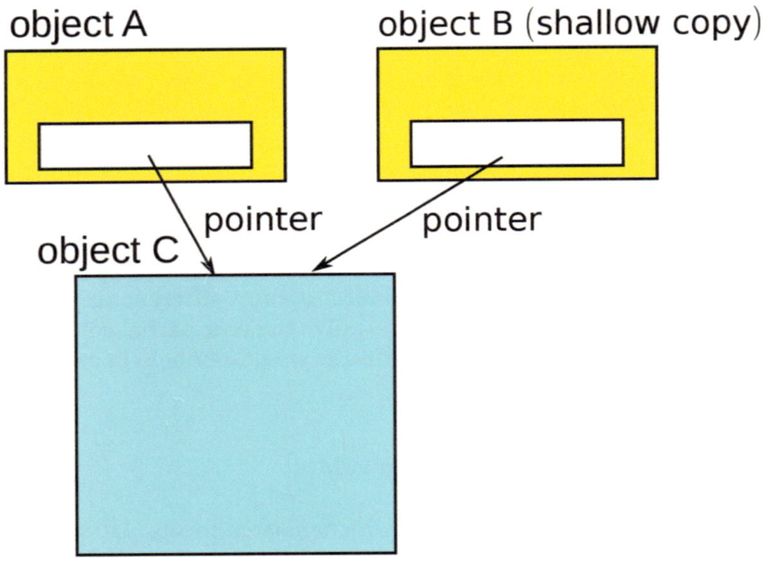

Figure 3.
An example of a shallow copy.

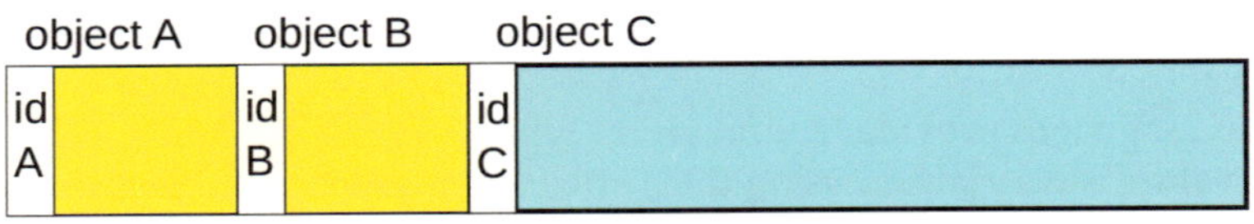

Figure 4.
Shallow copy serialization example.

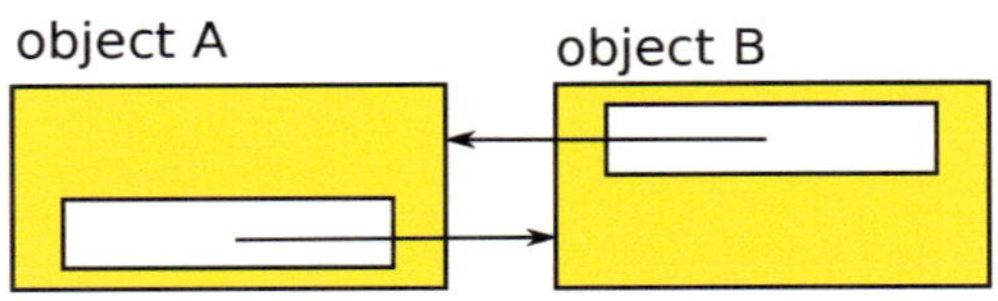

Figure 5.
Recursive object connection.

2.6 Inheritance and polymorphism

The object-oriented languages allow to reference an object by parent class pointer. During marshaling the complete object referenced by pointer should be stored, not the base class used as type of the pointer. This requires access to complete class inheritance hierarchy and runtime type information. The issue here is to retrieve and store unique and portable type identifier (e.g. C++ runtime type information is not portable) and use it to choose proper marshaling and unmarshalling procedures.

Inheritance becomes more troublesome when multiple inheritance is allowed, like in C++, in contrast to languages that permit only multiple interfaces (C#, Java). In such case the so-called diamond inheritance could be created, as depicted in **Figure 6**, where the class inherits from at least two different classes that have the same base class.

If *virtual inheritance* is used, only one instance of base class data is part of the final object; otherwise base class data is present multiple times as part of each class'

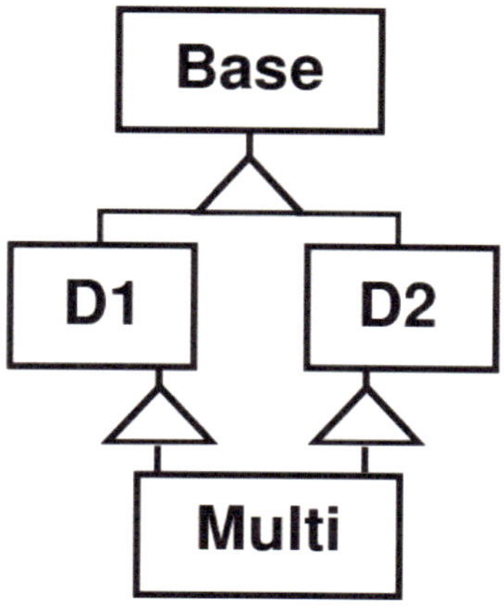

Figure 6.
Diamond inheritance class diagram.

```
class Base {};
class D1 : virtual public Base { };
class D2 : virtual public Base { };
class Multi: public D1, public D2 { };
```

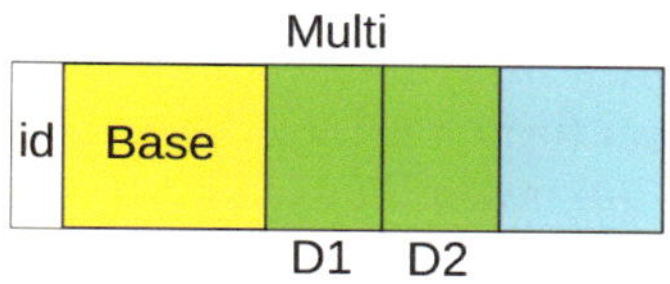

Figure 7.
Virtual diamond inheritance object serialization.

```
class Base {};
class D1 : public Base { };
class D2 : public Base { };
class Multi: public D1, public D2 { };
```

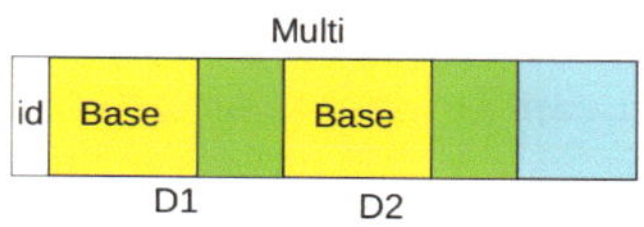

Figure 8.
Non-virtual diamond inheritance object serialization.

parents. The mechanism of serialization must detect which form of inheritance is used and serialize only one copy of the base class in case of virtual inheritance, as depicted in **Figure** 7, or save as many distinct versions of base class data as necessary, as shown in **Figure 8**.

2.7 Collections

It is quite common for serializable structures to contain some sort of data collections, like lists, dictionaries or sets. They do not provide much new issues for serialization process—at least when previously mentioned, serialization problems are solved in the given solution, but they might introduce discrepancies on semantic levels when porting data to different platforms. For example, default standard dictionaries in Java, C# and Python are *hash*-based, but in C++, before C++11 standard, only `std::map`, an ordered dictionary was available. In some extreme situations, it could lead to data that could not be unmarshalled on C++ side, when dictionary keys did not have natural ordering. On the other hand, data serialized in C++ could lose some of its information on transition to Python or would require usage of non-standard structures. Although most popular languages have similar set of basic collections, some subtle differences might lead to some semantics being 'lost in translation'. Either serialization tool provides its own implementation of common collections for each supported language, which might prove inconvenient for users as it creates impact on existing code of the applications, or serialization library has to introduce some common subset of requirements for collections and somehow verify those requirements before serializing data, which still might introduce some inconvenience for the users, but at least they can still use default collection types in their languages.

2.8 Backward compatibility

The software is constantly modified, and it is useful if the new version of software is able to read data saved by the older version. The backward compatibility can be achieved by storing archive version into stream and keeping deserializing code responsible for handling older versions. That leads to the newest code being the most littered with code for supporting previous versions, but such solution is probably the least memory consuming. Another approach is to use tagged fields, which also help with forward compatibility, as described below.

2.9 Forward compatibility

Forward compatibility allows to use older version of modules when new version of data format is introduced. Supporting forward compatibility requires ability to skip unknown fields in input data. It is usually achieved by adding tags—unique identifiers and type information—to each field. This way during deserialization the software can read only those fields it is aware of. Such approach provides some additional memory footprint, but helps to maintain software in live environment where upgrades are not possible to be performed at once or at all on some instances. Still this solution does not fix all issues—not every type of change in future data format can be made. For example, consider removing field—question arises how older software will interpret missing data, which it still expects. Adding new field should always work, but adding a new type might be troublesome in languages which do not have runtime reflection and type definition capabilities (like C++). Keeping software forward compatible remains mostly the developer's responsibility.

3. Overview of the existing tools

3.1 Java object serialization

One of the probably most known examples of serialization support built-in into language is Java object serialization [5]. Thanks to additional abstraction layer provided by Java virtual machine (JVM), serialization procedures do not need to be concerned with execution environment architecture—memory model, including endianness and object representation, is fixed by virtual machine. That solves some problems with portability of the archive between various real machines. The user only has to mark object using `Serializable` interface and pass object instance to data stream, which uses runtime object reflection to determine object's contents and properly serialize them.

Unfortunately such solution requires all applications involved in marshaling and unmarshalling to be running in JVM, so it is not a real cross-platform portability, but one limited to Java-related ecosystem. Additionally definition of objects has to be shared as parts of the code between applications, with limited support for forward and backward compatibility. Yet out-of-the-box availability and simplicity of use make such solution a good option for the homogeneous systems. Various other platforms implement similar solutions, including. NET `BinaryFormatter` [6] and Python `pickle` [7].

3.2 Google protocol buffers and apache thrift

The common approach for solving platform dependency issues is to introduce *interface description language (IDL)* which allows users to describe data using generalized rules. Such language usually provides limited capabilities, but it makes possible to generate proper code for various platforms. For example, both Google Protocol Buffers [8] and Apache Thrift [9] allow basic numeric types, enumerations, lists, and dictionaries, but no polymorphism or object references. In return those tools can generate serializers and deserializers for significant range of languages, starting with C++, through C# and Java, to Python and JavaScript. Users gain portability in exchange for enforced data structures.

Apache Thrift can serialize objects described with common IDL using various target methods, including human-readable JSON, but for highest efficiency the so-called Compact Protocol should be used, which is similar to the serializer present in Protocol Buffer. Using overhead of field identifier and type encoded before each value, those encoders ensure good forward and backward compatibility. Archive users still need to share part of the code, in the form of the IDL files, but can be implemented in various technologies. Due to built-in forward compatibility support, there is also a smaller chance that change in the IDL would need to propagate to all involved parties than previous solutions.

Some solutions try to remove field identifier overhead from archive and keep forward compatibility support by including whole schema (IDL) inside archive [10, 11]. It can significantly reduce archive size when storing multiple items of the same type. Initial parsing of the schema can introduce some processing overhead, but more importantly such solution might be inconvenient for languages with static typing, where using types created in runtime might be tiresome for developers.

3.3 C++ dedicated solutions

Although C++ is usually supported by cross-language solutions, like Apache Thrift or Google Protocol Buffers, it lacks its own in-language serialization support [12], like Java, C# or Python. Using IDL-based serialization is not always an option for C++ project, as it can be less efficient or too limited than language-specific solution. It is also required to generate all data structures from IDL, so it is not suitable to use in existing project, where serialization should be added to legacy code. As C++ is often used in big legacy projects, the need for language-specific serialization library is justified.

3.3.1 C++ Boost.Serialization

`Boost.Serialization` [13] is a widely used C++ serialization library. This library makes reversible deconstruction of an arbitrary set of C++ data structures possible. It uses only C++03 facilities [14] and therefore could be used in a wide spectrum of C++ compilers, even on the tools that do not support newer C++11 standard [15]. The archive could be binary or raw text or XML. The serialization does not require external record description; additionally the user types to be serialized do not need to derive from a specific base class.

The library supports writing and reading of built-in types (numbers) and most classes from the standard library, like `std::string` and other containers. Enumerations are saved as numeric values. `Boost.Serialization` supports pointer and reference marshaling and demarshaling, i.e. the serialization acts also for the data pointed to. Additionally this library has support for shared pointers (only one copy of data pointed to is saved) and objects with multiple inheritance (also virtual inheritance). Unfortunately, the Boost.Serialization has no forward compatibility. Portability between platforms is achieved only for text formats. In binary format only the data is stored, without additional information that allows the type identification. The numbers are stored as their memory image. The archive is very efficient in terms of size; processes of reading and writing are fast. Unfortunately, there is a lack of portability of binary format.

The backward compatibility is achieved by adding the versioning; therefore the library user can provide different solutions for each version.

To add the serialization for user type, the programmer should implement method serialize, where one of its argument is archive and the second is the version number.

3.3.2 C++ cereal

Similarly to Boost.Serialization, cereal library [16] provides language-specific serialization capabilities. It too makes developer responsible for writing explicit marshaling procedures and supports only backward compatibility. Because it requires serialized structures to use C++11 smart pointers instead of raw pointers and references, library's code can be much simplified, and object tracking becomes easier. Additionally cereal provides support for most of C++ standard library, making it more convenient than Boost.Serialization.

Library contains similar archive types as Boost.Serialization including JSON, XML and binary formats. The major advantage of cereal is the presence of PortableBinary archive, which allows to store data efficiently in binary format in a cross-platform way—Boost.Serialization does not support moving archive across platforms with various endianness, although it is currently being developed by Boost team.

4. cereal_fwd: New serialization library for C++

4.1 Motivation

Currently C++ ecosystem seems to lack efficient and convenient serializing tool supporting portability and forward compatibility. Developers can use some of cross-language tools, like Apache Thrift or Protocol Buffers, but those enforce data types used in application. Users that want to add serialization to existing code base, with already defined C++ classes, are usually left with Boost.Serialization (as part of the most popular Boost library), whose binary archives are not portable. Other solutions, like cereal, do not provide forward compatibility support. The lack of desired tool was a motivation to create new library for serialization of C++ objects.

The main goals for the new C++ library were:

- Support backward and forward compatibility.
- Use minimal size of saved data without hindering ability to evolve structure of serialized data.
- Support streaming for saving and loading operations.
- Minimize allocations during saving and loading process.
- Support portability between different platforms.

- Do not break compatibility for existing archives in library.
- Loading data from unknown source cannot result in undefined behaviour.

4.2 Implementation

New cereal_fwd library was based on cereal as it already provides some of the required features, and thanks to relying on C++11 language features, it has much simpler implementation than popular Boost.Serialization. It might be troublesome for some users, as it requires code to be C++11 compliant, but it helps keep library code simple, and transition towards C++11 should be desired by most existing code's maintainers anyway.

Users of both cereal and cereal_fwd are required to explicitly list serializable structure's fields in a dedicated method, as shown in **Figure 9**. This is a similar solution to the Boost.Serialization, as as of the moment C++ lacks any reflection support, which could help automate the process. Ongoing work [17] suggests that in the future such code could be significantly simplified.

Figure 10 shows how structure can be serialized using selected archive type. In the example BinaryArchive is used; cereal provides few more archive types, each with the same interface—choosing different archives does not require any changes on structure side. One of the existing archives—PortableBinary—provides support for platform portability and was used as a starting point for cereal_fwd extension in the form of the ExtendableBinary archive. The new archive type is responsible for supporting forward compatibility in cereal_fwd.

4.2.1 Numbers

Numbers in cereal_fwd are serialized and deserialized similar to cereal, but to ensure forward compatibility, all integer type fields are saved with size information attached. This makes it possible to load integers which have different sizes on writing and reading side. The attempt to load integer that cannot fit into type used for loading will result in thrown exception. For space-saving purpose, the size of the saved integer is determined as the minimal number of bytes needed to represent the number being stored.

```
struct Example
{
  std::uint8_t x;
  float y;

  template <class Archive>
  void serialize(Archive& ar)
  {
    ar(x, y);
  }
};
```

Figure 9.
Basic cereal *serialization procedure.*

```
int main()
{
  std::ofstream os("out", std::ios::binary);
  cereal::BinaryOutputArchive archive(os);

  Example data;
  archive(data);

  return 0;
}
```

Figure 10.
Serializing structure in cereal.

4.2.2 Arrays and strings

The size of array or string is saved using variable-length integer encoding—the most significant bit is used to indicate if the next byte is part of stored number. This way for small arrays size is stored using only 1 byte. To speed up serialization and save archive space, all items in the array are assumed to have the same size—size information is stored only once per array.

4.2.3 Polymorphic types

To properly read object stored using polymorphic pointer, identifier of the object's most derived type is needed. In cereal_fwd library, as well as in Boost.Serialization, for polymorphic pointers a unique type of identifier is saved to stream. Identifier can be any string; by default fully qualified name of the type is used. During reading process identifier is a key in factory which helps in creating correct object and selecting proper deserialization function.

Identifier for specific class is saved only once in stream, for the first occurrence of given type, accompanied with corresponding ordinal number. For every next instance, just the ordinal number is saved. While reading data saved by newer application, it may happen that identifier of polymorphic type will be connected to field which is unknown and will not be normally read. For that reason logic to process every occurrence of type identifier was added.

During code evolution new derived types for existing base classes may be added and saved as polymorphic pointers in archive. In the original version of cereal library, the attempt to read unknown polymorphic type results in exception being thrown, consequently stopping reading process. In cereal_fwd library an option was added which changes that behaviour. For pointers of unknown type, nullptr value is set, and reading process is continued without interruption.

4.2.4 Shared pointers

Existing cereal library supports saving several pointers pointing to the same object. In this situation only one copy of the data is saved into the stream. For every other occurrence of the same object, only numerical identifier of previously saved data is stored. For saving process a dictionary of stored pointers and their identifiers is maintained. If address is found in the dictionary, only identifier is saved. For loading process similar mapping between identifiers and shared pointers is maintained. If during loading pointer identifier is already in the map, data is not loaded, and pointer is returned directly from the map.

4.3 Forward compatibility

Forward compatibility was a desired important new feature of cereal_fwd, yet implementing it proved to be a demanding task. Some initial assumptions and design decisions were challenged during implementation and are described below.

4.3.1 Adding fields

Supporting possibility for adding fields in newer versions of application should be straightforward—old version of application should just ignore unknown fields. It is true, as long as shared objects are not concerned. Class modification may lead to a new shared pointer field being added, which points to an object which is already used by a different class in the older version of the application. If the first occurrence of shared pointer is saved by field which is present only in the new version and the older version used for reading, reading such shared pointer may be difficult. An example of such situation is depicted in **Figure 11**. Object saved in the archive is A class. In the first version of the application, only one pointer to C class object is saved. In the second version, B::c field pointing to C class object is added. Fields A::c and B::c point to the same object. In the second version, B::c pointer, being part of b field of class A, is saved first. Next, c field of A class is saved. When data is read by the first version of the application, during reading of the B::c, the type of that field is not known, and the whole field could be skipped in basic situation. However, data has to be read in some way to be available for A::c field restoration.

One of the solutions to the described problem is saving stream position of each occurrence of shared pointers and restoring it in case data is needed to read the object by other pointers. This solution was rejected because it would introduce additional requirement on streams supported by the library, to handle rewinding reading position. Such operation is not supported by all streams, i.e. streams formed using network sockets. Another rejected solution was partial interpretation of data and storing it in temporary objects of basic types supported by archive. If the object pointed by shared pointer from unknown field needs to be read by other pointers, instead of using main stream, data would be copied from temporary objects. This

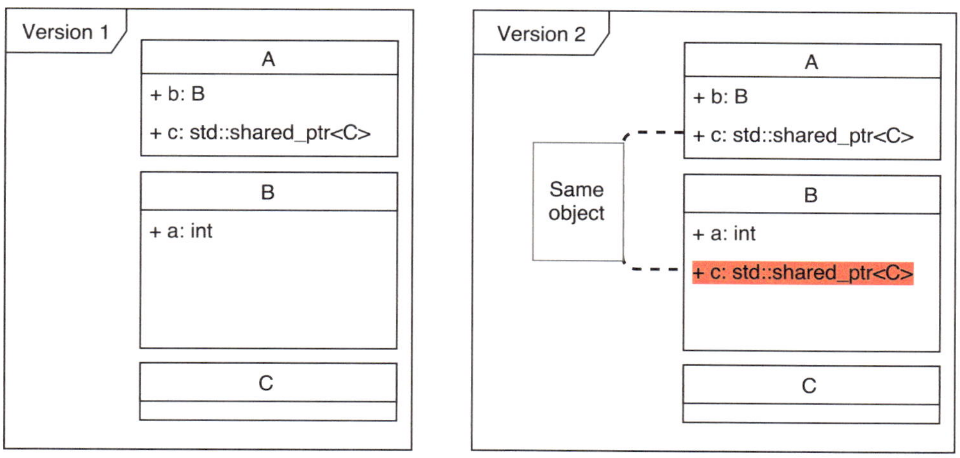

Figure 11.
Example of class evolution showing problem with saving shared pointers.

approach would introduce computational overhead even if no other pointer to the same object was saved.

The chosen solution copies binary data of pointed object of unknown field type to a temporary buffer, by default allocated on heap. In case data needs to be read for earlier omitted pointer, it is read from helper stream created from buffer. Apart from additional stream, stack of reading positions is maintained. It is needed in case omitted object contains additional pointers which were also omitted. An example of this case can be seen in **Figure 12**. The main saved class is `Outer`. In the first version of application, two shared pointers `Outer::q` and `Q::z` are saved. In the second version, pointers are saved in the following order: `Inner::z`, `Inner::q`, `Q::z`, `Outer::q`, `Q::z`. Only single object instances of `Q` and `Z` classes are stored; pointers point to the same objects. When data saved by the second version is read by the first one, data needed by `Outer::q` field can be found in `Inner::q` position. Data for `Q::z` field can be found in `Inner::z` position.

Making copy of data may require a lot of memory. In a worst case size of copied data may be close to the size of all data being read. Such memory allocations may not be acceptable in some applications. Because of that, option to limit maximal size of helper buffer has been added. Trying to use more memory will result in exception being thrown.

4.3.2 Removing fields

Apart from adding new fields, at some point of application evolution, it might be justified to remove no longer needed fields. Saving unnecessary fields results in size and computational overhead. Because in `cereal` order of saving fields is used for field identification, it is forbidden in that library to erase fields from saving and reading methods.

To circumvent this problem, `cereal_fwd` adds ability to change field's type to `OmittedFieldTag`. Using this type indicates position where there was a field but it is no longer used. The older version of application, when trying to read field, which in archive is marked using this tag, will just leave field value in the object unchanged. In most situations it will leave that field with default value assigned in object constructor. It is the responsibility of the developer to erase only such fields

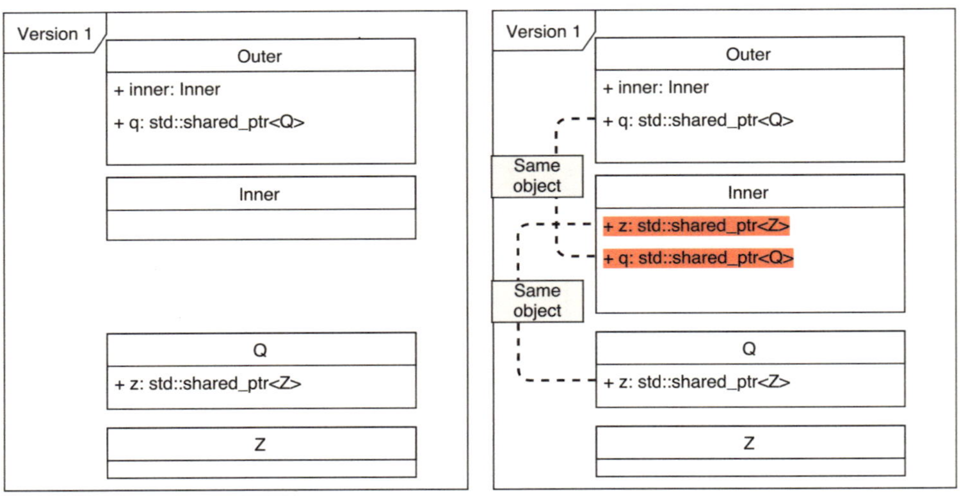

Figure 12.
Example of nested shared pointers.

in the newer application, where default values will still make older versions of the application work correctly. To help the developer create a robust code, method `Archive::wasSerialized` was added. Result of this method indicates if the field being read was really saved or if `OmittedFieldTag` was used in the archive.

4.3.3 `cereal_fwd` *forward compatibility summary*

The `cereal_fwd` supports forward and backward compatibility. The forward compatibility is available, for the following changes:

- Introducing version parameter in `serialize` method, which can be used for conditional serialization code.
- Adding new fields at the end of the object's serialization code; new fields have to be loaded conditionally using class version stored in archive.
- Removing fields from the end of class/struct is permitted.
- Changing the size of integers, as long as stored value is not bigger than target field size; exception is thrown otherwise.
- Renaming structures or classes; the exceptions, caused by storing fully qualified type name in the archive:
- Renamed class cannot be stored using shared pointer.
- Renamed class cannot be saved using pointer to the base class.
- Changing type of serialized field to `OmittedFieldTag`; this change can be done even without changing class version number and introducing conditionals to serialization procedure.
- This leaves some changes in the data structure layout to be still forward incompatible.
- Removing fields without adding `OmittedFieldTag`—trying to load more fields than were saved will result in exception.
- Changing the sign of an integer and loading number that does not fit into a new type, e.g. loading negative number to unsigned integer type.
- Changing the size of floating-point types (e.g. from float into double).
- Changing order in which fields are serialized.
- Adding field for serialization without updating class version and loading it without checking version.
- Adding field for serialization not as the last field.

It is still the developer's responsibility to introduce changes in such a way that archives will be forward compatible.

4.4 Library benchmarks

Performance and memory consumption of newly created `cereal_fwd` was compared against original `cereal`, `Boost.Serialization`, as similar library, and Protocol Buffers as popular non-C++-dedicated solution. The time taken to serialize and deserialize data was comparable; however the fastest for numbers and collection of numbers was Protocol Buffers; the slowest turned out to be `Boost.Serialization`. For pointers the fastest was `cereal` or `cereal_fwd`, the slowest was `Boost.Serialization`, and Protocol Buffers does not support deep pointers serialization.

Usage of memory allocated on heap was measured using total number of allocations (number of calls to memory manager) and maximal size of the heap. All libraries (`Boost.Serialization`, Protocol Buffers, `cereal` and our `cereal_fwd`) had similar usage of the memory when serializing numbers, collections and pointers; all differences were not statistically significant.

Size of created executable was 30% bigger for our `cereal_fwd` than for `cereal` and comparable with `Boost.Serialization`. Protocol Buffers had the smallest code size for serializing numbers, but in the case of collections, code size was the biggest.

The benchmark results are available at project web site.

5. Discussion and conclusion

The support for serialization is required in all currently used programming languages, because there still a growing need to exchange information [18]. The perfect solution, meeting all requirements, does not exist, because the requirements are contradictory:

- The fastest serialization/deserialization process is achieved for binary format, but it is not portable between platforms.
- The most compact is also binary format (without included data description), but such archive is hard to use to exchange information between modules written in different programming languages.
- The text formats, like JSON or XML, are portable and self-descriptive, but serialization/deserialization needs additional data processing, and archive takes significantly more space than the binary one and in result might be slower to transmit if needed.

The programming languages that support reflection have simplified serialize/deserialize process, but other environments needing several technical issues should be resolved, as depicted in Section 2. The existing libraries to serialization require constant development and modernization; because the programming languages are modernized, new standards are implemented, and additional new programming languages are being developed and become used. Our new `cereal_fwd` library addressed the forward compatibility problem for C++ serialization. The library is publicly accessible at `https://github.com/breiker/cereal_fwd` under BSD-like licence.

Acknowledgements

This work was supported by the Statutory Founds of Institute of Computer Science.

Competing interests

The authors declare that they have no competing interests.

Availability of supporting data

Supporting data, including examples of use and source codes, is available in the project repository `https://github.com/breiker/cereal_fwd`.

Author details

Konrad Grochowski, Michał Breiter and Robert Nowak*
Institute of Computer Science, Warsaw University of Technology, Poland

*Address all correspondence to: robert.nowak@pw.edu.pl

References

[1] Sumaray A, Kami Makki S. A comparison of data serialization formats for optimal efficiency on a mobile platform. In: Proceedings of the 6th International Conference on Ubiquitous Information Management and Communication. ACM; 2012. p. 48

[2] Keith W Ballinger, Erik B Christensen, and Stefan H Pharies. Xml serialization and deserialization. US Patent 6,898,604; 2005

[3] IEEE. IEEE Standard for Floating-Point Arithmetic. IEEE Std 754-20082008. pp. 1-70

[4] International Organization for Standardization. Unicode. ISO/IEC 10646; 2008

[5] Oracle. Java Object Serialization Specification. 2019. Available from: https://docs.oracle.com/en/java/javase/12/docs/specs/serialization [Accessed: April 15, 2019]

[6] Microsoft. BinaryFormatter Class. 2019. Available from: https://docs.microsoft.com/en-us/dotnet/api/system.runtime.serialization.formatters.binary.binaryformatter?view=netframework-4.7.2 [Accessed: April 15, 2019]

[7] Python Software Foundation. Pickle—Python Object Serialization. 2019. Available from: https://docs.python.org/3/library/pickle.html [Accessed: April 15, 2019]

[8] Google Inc. Protocol Buffers–Google's Data Interchange Format. 2019. Available from: https://github.com/protocolbuffers/protobuf [Accessed: April 15, 2019]

[9] Apache Foundation. Apache Thrift. 2019. Available from: https://thrift.apache.org [Accessed: April 15, 2019]

[10] Apache Foundation. Apache Avro. 2019. Avaliable from: https://thrift.apache.org [Accessed: April 15, 2019]

[11] Huang AS, Olson E, Moore DC. Lcm: Lightweight communications and marshalling. In: 2010 IEEE/RSJ International Conference on Intelligent Robots and Systems. IEEE; 2010. pp. 4057-4062

[12] Nowak R, Pajak A. C++ Language: Mechanisms, Design Patterns, Libraries. BTC, Legionowo. ISBN: 978-83-60233-66-5; 2010

[13] Boost Community. Boost. Serialization. 2019. Available from: https://www.boost.org [Accessed: April 15, 2019]

[14] International Organization for Standardization. Programming Languages—C++. ISO/IEC 14882:2003, 2003

[15] International Organization for Standardization. Programming Languages—C++. ISO/IEC 14882:2011, 2011.

[16] Shane Grant W, Voorhies R. Cereal. 2019. Available from: http://uscilab.github.io/cereal/ [Accessed: April 15, 2019]

[17] International Organization for Standardization. Working Draft, C++ Extensions for Reflection. 2019. Available from: http://www.open-std.org/jtc1/sc22/wg21/docs/papers/2018/n4766.pdf [Accessed: April 15, 2019]

[18] Malladi SK, Murphy RF, and Deng W. System and method for processing messages using native data serialization/deserialization in a service-oriented pipeline architecture. US Patent 8,763,008; 2014